ADR and Adjudication
in Construction Disputes

ADR and Adjudication
in Construction Disputes

Peter Hibberd
MSc, FRICS, ACIArb
and
Paul Newman
MA (Cantab), DipLaw, FCIArb, Barrister

**Blackwell
Science**

© by P. Hibberd and P. Newman 1999

Blackwell Science Ltd
Editorial Offices:
Osney Mead, Oxford OX2 0EL
25 John Street, London WC1N 2BL
23 Ainslie Place, Edinburgh EH3 6AJ
350 Main Street, Malden
  MA 02148 5018, USA
54 University Street, Carlton
  Victoria 3053, Australia
10, rue Casimir Delavigne
  75006 Paris, France

Other Editorial Offices:

Blackwell Wissenschafts-Verlag GmbH
Kurfürstendamm 57
10707 Berlin, Germany

Blackwell Science KK
MG Kodenmacho Building
7–10 Kodenmacho Nihombashi
Chuo-ku, Tokyo 104, Japan

First published 1999

Set in 10/12pt Palatino
by DP Photosetting, Aylesbury, Bucks
Printed and bound in Great Britain by
MPG Books Ltd, Bodmin, Cornwall

The Blackwell Science logo is a trade mark of
Blackwell Science Ltd, registered at the United
Kingdom Trade Marks Registry

DISTRIBUTORS

Marston Book Services Ltd
PO Box 269
Abingdon
Oxon OX14 4YN
(*Orders:* Tel: 01235 465500
         Fax: 01235 465555)

USA
Blackwell Science, Inc.
Commerce Place
350 Main Street
Malden, MA 02148 5018
(*Orders:* Tel: 800 759 6102
              781 388 8250
         Fax: 781 388 8255)

Canada
Login Brothers Book Company
324 Saulteaux Crescent
Winnipeg, Manitoba R3J 3T2
(*Orders:* Tel: 204 837 2987
         Fax: 204 837 3116)

Australia
Blackwell Science Pty Ltd
54 University Street
Carlton, Victoria 3053
(*Orders:* Tel: 03 9347 0300
         Fax: 03 9347 5001)

A catalogue record for this title
is available from the British Library

ISBN 0-632-03817-9

Library of Congress
Cataloging-in-Publication Data
is available

For further information on
Blackwell Science, visit our website:
www.blackwell-science.com

Contents

Preface

Over the past thirty years or so construction disputes have generated increased interest. Once, the engineer or architect resolved virtually all claims under the contract and dispensed their own form of justice. The parties generally accepted these rulings but increasingly decisions came to be seen as challengeable. Traditionally, disputes were litigated but the system's failure to provide satisfaction led to the growth of arbitration to settle construction disputes.

The advent of general contractors outsourcing much of the construction process to subcontractors created another source of dispute, particularly about payment. Arbitration proved ineffective for such disputes and adjudication and other methods emerged as an alternative. Consequently this search, together with the upbeat marketing of alternative dispute resolution (ADR), meant that conciliation, mediation, MedArb, mini-trial, dispute review boards and other techniques became topical. Many of these concepts originated in the United States and outside construction. Interest then spread to the United Kingdom and elsewhere. This growth in the use of alternative dispute resolution techniques prompted us to write this book.

Our aim was to describe the processes of ADR and adjudication, warts and all, to provide a basis for an informed choice of dispute resolution and the conduct of the various processes. Naturally, we have had to touch briefly on arbitration and litigation while doing this.

The book is intended to meet the needs of practitioners and contracting parties, while retaining an interest for both students of ADR and academics. Coverage of the practices of ADR, as well as events leading up to its introduction, are dealt with. The most important of the standard forms of contract are included, so as to provide an overview of approaches as well as the mechanics of these forms.

Inevitably, the introduction of the Housing Grants, Construction and Regeneration Act 1996 (HGCRA) and the Arbitration Act 1996 had considerable impact on the book. These, together with the Scheme for Construction Contracts, the Woolf Reforms and the Construction Industry Model Arbitration Rules (CIMAR) have meant that dispute resolution is now a topic of much interest. The HGCRA introduced a statutory form of adjudication and a number of the contract authoring bodies produced schemes covering adjudication procedure and also revised their conciliation procedures in their standard forms.

The book is based on the law as at January 1999, save for the following important reference to the cases of *Macob Civil Engineering Ltd* (1999) v.

Morrison Construction Ltd 16-CLD-05-06, *Outwing Construction Ltd* (1999) v. *H. Randall & Son Ltd* 16-CLD-02-01 referred to below. Appendix 5 briefly outlines the Woolf Reforms and the Civil Procedure Rules operative from April 1999.

As we discuss in Chapter 9, adjudication under the HGCRA not only addresses a problem but will bring with it many other problems. The first of such matters found its way to the Technology and Construction Court, where Mr Justice Dyson heard the case between *Macob Civil Engineering* v. *Morrison Construction*. This case concerned the status of an adjudicator's decision under the Scheme for Construction Contracts (England and Wales) Regulations where such a decision had been challenged. The judge decided that:

(1)   The decision of an adjudicator made in accordance with the Scheme was binding on the defendant until the dispute arising from the decision was finally determined by arbitration, legal proceedings or agreement.
(2)   The defendant was required by that decision to pay the sums identified by the adjudicator forthwith.

The contract between the parties fell under the ambit of the HGCRA but, as it did not comply with its requirements, the provisions of the Scheme applied. The adjudicator, in giving his decision, made an order under paragraph 23 (1) of the Scheme that the parties comply peremptorily with his decision and gave permission under section 42 of the Arbitration Act 1996 to enforce the decision. The enforcement of the order was resisted on the ground that the adjudicator must give a valid decision, and one that breached the natural rules of justice was not a valid decision and was therefore not binding or enforceable as a contractual term. Mr Justice Dyson rejected this argument on the grounds that it would be all too easy to challenge an adjudicator's decision as a breach of natural justice, and if this was permissible, it would substantially undermine the effectiveness of the Scheme set up by the Regulations.

A further argument offered on Morrison's behalf was that the court had no power to make an order under section 42 of the Arbitration Act 1996, as modified by the Schedule to the Regulations, because this power was exercisable 'unless otherwise agreed by the parties' and a clause in a contract that referred to a dispute arising out of an adjudicator's decision to arbitration was such an agreement. Mr Justice Dyson also rejected this argument. Leave to appeal was given, so the story is unlikely to end here.

Following fast on this case was *Outwing* v. *H Randall & Son* where commentators have said that this case goes even further than the *Macob* decision in that the courts will take steps to ensure speedy enforcement of adjudicator's decisions.

The dispute resolution topography is now extremely cluttered. ADR and adjudication may challenge the traditional supremacy of arbitration and litigation particularly if parties believe the quick fix more desirable than slow, painful warfare. Anecdotal evidence suggests adjudication is becoming more common after a slow beginning. ADR still has the whiff of fringe medicine and has not become part of the corporate psyche yet. The CPR may change this but it is questionable the judiciary has itself developed a sufficient focus to spearhead any surge forward for ADR.

Clearly, many problems will continue to occur and ADR is a means of resolving such difficulties.

The observations on the development of ADR in Australia were drafted by Ian Briggs of Minter Ellison, Brisbane, to whom the authors are grateful for his assistance.

Peter Hibberd
Paul Newman
June 1999

List of Abbreviations

AAA – American Arbitration Association
BEC – Building Employers Confederation
Blue Form – BEC Domestic Subcontract
BPF – British Property Federation
CASEC – Committee of Associations of Specialist Engineering Contractors
CDRG – Construction Disputes Resolution Group
CEDR – Centre for Dispute Resolution
CIArb – Chartered Institute of Arbitrators
CIC – Construction Industry Council
CIMAR – Construction Industry Model Arbitration Rules
CPR – Civil Procedure Rules
DOM/1, DOM/2 – Standard Form of Subcontract for Domestic Sub-contractors
DRA – Dispute Resolution Adviser
DRB – Dispute Review Board
FASS – Federation of Specialist Subcontractors
FIDIC – Fédération Internationale des Ingénieurs-Conseils
GC/Works/1 – General Conditions of Contract for Building and Civil Engineering Major Works
Green Form – CASEC Nominated Subcontract
HGCRA – Housing, Grants, Construction and Regeneration Act 1996
ICC – International Chamber of Commerce
ICE – Institution of Civil Engineers
IChemE – Institution of Chemical Engineers
JCT – Joint Contracts Tribunal
NAM/SC – JCT Standard Form of Named Subcontract Conditions
NEC – New Engineering Contract
NFBTE – National Federation of Building Trades Employers
NJCC – National Joint Consultative Committee for Building
NSC/A – JCT Nominated Subcontract Agreement
NSC/C – JCT Nominated Subcontract Conditions
NSC/T – JCT Nominated Subcontract Tender
NSC/3 – JCT Standard Form of Nominations of Subcontractor
NSC/4 – JCT Standard Form of Nominated Subcontract NSC/4
ORSA – Official Referees and Solicitors Association (now TecSA – Technology and Construction Court Solicitors Association)
PM – Project Manager

PSA/1 – General Conditions of Contract for Building and Civil Engineering Works

RIBA – Royal Institute of British Architects

RICS – Royal Institution of Chartered Surveyors

RSC – Rules of the Supreme Court

The Scheme – The Scheme for Construction Contracts Regulations

SFBC – Standard Form of Building Contract

UMIST – University of Manchester Institute of Science and Technology

UNCITRAL – United Nations Commission on International Trade Law

Chapter 1
Construction Contracts: An Inevitable Source of Conflict?

Conflict, claims and disputes

Many involved in the construction industry believe that conflict is inevitable and that this is promoted by the standard forms of contract used. However, it is, to the authors, axiomatic that the potential for conflict is inherent in any contractual situation. Although both parties have the same objective – a building – the motivation for providing the building is generally, but not always, concerned with the profit to be earned (or benefit to be obtained). Profit for the contractor and for the promoter of the building is earned in different ways. It is often assumed this means that when the profit position of either party is threatened, conflict will emerge. But is this true and, if it is, is the conflict caused or worsened by the standard forms of contract that are used?

The potential for conflicts, disputes and claims always exists and it is beneficial to recognise this from the beginning of any contractual relationship. It is also worth considering the distinction between these terms in a little more detail before proceeding, not in depth but simply to provide a basis of understanding, in the belief that to do so will reduce the need for litigation and arbitration.

There is to our mind a significant difference between simply exercising one's rights under a contract (a claim) and the existence of a dispute. A claim is no more than an assertion and cannot become a dispute until there is a genuinely disputable issue. This was the view expressed in *Mayer Newman & Co. Ltd* v. *Al Ferro Commodities Corporation SA* (1990) and although this was in the context of arbitration, it does seem to be a view worthy of wider application. This view would also accord with comment made in *M.J. Gleeson Group* v. *Wyatt Snetterton Ltd* (1994), and the case of *Halki Shipping Corporation* v. *Sopex Oils Limited* (1996) is also illustrative of the need to know whether a dispute exists. There is an even more significant difference between a dispute and a conflict, although not everyone would agree on the precise nature of the difference; conflict itself cannot necessarily be determined by the dispute resolution process within the contract.

It is submitted that a dispute, by definition, arises if there is a genuine difference of opinion regarding the interpretation and implementation of the contract. This corresponds to the dictionary definition 'to make the subject of argument, to contend for, to oppose by argument, to call into question' and also accords with *Hayter* v. *Nelson and Home Insurance Co.* (1990) whereby the word 'disputes' in an arbitration clause has its ordinary meaning. Nevertheless, some arbitration agreements and some contract provisions make reference to both disputes and differences being referable, in order to ensure the matter can be determined by the forum. While any difference of view is respected by the parties, no conflict exists, simply a dispute. Naturally, conflict can follow and this depends on the behaviour of those involved in the dispute. Conflict here is also being given its dictionary meaning and is where some real element of tension exists.

However, it is recognised that the distinction we make between disputes and conflict is not universally held. Brown and Marriott[1] discuss this and draw on definitions provided by Mediation UK and others[2], which express conflict rather differently. To them, a dispute is one type of conflict.

These various definitions are fundamental to understanding the best way of tackling the problem. Brown and Marriott state that, 'The distinction between behavioural conflict and justiciable dispute, even though this may not always be clear, is relevant to Alternative Dispute Resolution (ADR) because of the different approaches which may have to be adopted...'. It seems to us that dispute procedures operate best when little or no conflict exists. Once conflict emerges, the solution to the problem becomes quite different and it is for this reason we recognise a distinction. In determining what process should be adopted to find a solution to a problem that has arisen on a project, one must first establish the nature of the problem.

Problems vary appreciably and may range from one that requires a technical solution to one where working relationships have broken down. The latter may be a result of the former and because it has been badly handled. Each time a problem occurs it is essential that one identifies its nature and then adopts the most appropriate process of resolution. The time between the emergence of a problem, the adoption of a resolution and or conflict is extremely variable. Occasionally, because of behavioural characteristics, conflict arises even before dispute resolution mechanisms can be commenced. Clearly, it is imperative that those charged with settling contractual and project problems recognise these matters and the need for adopting different solutions to meet disparate problems; also, that they recognise the need to attempt to resolve disputes as soon as possible.

The problem of profit

The respective profit positions of a project may change as a result of the other party's action, and where this happens it may be reasonable to look to the contract to seek redress. The respective profit positions, also, may be threatened because of circumstances beyond the control of the parties to the contract. Again, they will look to the contract to secure some form of recompense.

This reaction to changing levels of profitably is considered by some observers to be short-sighted, primarily concerned with the current project. It is suggested by some within the industry that a longer-term view would not necessarily provoke this response. However, others feel that the longer-term view is to some extent irrelevant because of the tendering procedures adopted. Furthermore, regardless of short or long-term relationships, it is quite proper to look to the contract for redress if a situation emerges different from that contracted; after all, it is argued, that is the purpose of the contract.

Tendering procedures, which require each project to be tendered on initial price alone or where price is the paramount criterion on which selection is made, do not lend themselves to the development of a long-term business relationship. Notwithstanding the problems that arise, many people still advocate open competition or some form of restricted competition associated with a single stage tendering procedure. The fixed price lump sum contract has its attractions, especially for those who only build once or very occasionally. For these building clients the business relationship lasts little longer than the project itself, although the consequences may be felt for much longer. In these circumstances, it is perhaps not surprising and indeed natural, that the parties try to 'exploit' the contract terms. The pursuit of claims to improve margins following cut-throat tendering is a major contributor to the problems of the construction industry.

'Partnering', first introduced by the US Army Corps of Engineers, has been put forward as a way to change the culture and to reduce this type of problem. It was pioneered in order to resolve the problem of disputes arising from fixed price low bid public contracts. Although 'partnering' offers certain clear advantages, it also has disadvantages. The profit may be more easily protected over time but the opportunity to secure greater profit does not always operate equally between the parties. Furthermore, it also increases the risk of unethical practices and is not so obviously accountable. The Australian construction industry is experiencing such a problem and some speculate that procurers will resort to open tendering procedures. If this were to occur in the UK it would be seen as a retrograde step and it suggests that 'partnering' as a means of overcoming the disputes and conflict that arise from adopting short time horizons may be short-lived. At present 'partnering', prompted by Latham and more recently by Egan, is being viewed as a non-adversarial approach to contracting and a number of major clients are experimenting with a so called 'win-win' arrangement.

The contract as an agreement and a means of redress

As mentioned earlier, the parties' views of the contract may change out of economic necessity and as a consequence of the way the contract is procured. However, a party's view can also change for a variety of other reasons related to the project (as opposed to contract) procurement. The respective positions of the parties can change for a wide range of reasons, including variations required by the client; unforeseen ground conditions; strikes; inflation; unavailability of materials; weather; and circumstances caused by the incompetence of the contractor, consultants and even the promoter. Whether these changes should or will create a dispute and or conflict is another matter but there is evidence to suggest that they do on many projects.

Using the contract to seek redress is not that surprising and few would argue that this should not occur. The real issue is not whether the contract should be used, but rather in what circumstances and in what way it should be used. The contract is best seen, not so much as providing a means to facilitate the assertion of rights, but rather as a framework for providing a solution. If the parties' intentions are clear then any change could be regulated by the inclusion of appropriate provisions within the contract. There is absolutely nothing wrong with this in principle but seldom are intentions absolutely clear. Even where they are, the consequences of strict implementation of the contract provisions frequently produce an unacceptable position to one or other party.

In the context of professional services, it has been said[3] that the contract is the most important document for avoiding disputes, if it includes the necessary basic elements. This is equally applicable to construction contracts. However, one can argue that if a contract is properly considered and thus reduces disputes, then it may be equally true that inadequate consideration will more likely cause disputes. One can make an important distinction between avoiding disputes and causing them. Transaction cost theory helps us to understand this distinction and the appropriate level of preparation and consideration of information, prior to contract.

The contract itself cannot cause conflict, only the parties to the contract or their advisers can. In theory, the contract is or should be a reflection of the requirements of the contracting parties and therefore should not even produce disputes. It should of course provide a framework for the exercise of, and account for, the way risk has been apportioned. Herein lie several problems. Although in theory contracts reflect the requirements of the parties, they seldom do in practice. This happens partly because of the use of standard forms of contract as a shorthand means to arrive at a set of contract conditions, without adequately considering their actual terms and whether they fulfil requirements; and partly because of genuine problems of misunderstanding and misinterpretation and because of the changing perception of the participants, i.e. what they expect from the project and the contract changes throughout the project's life.

Our view that a contract cannot cause conflict is perhaps challenged by Clegg[4] in the title of his paper and by the words he uses: 'Contracts cause conflicts because they are the rational occasions whereby indexicality can be exploited by self interested professionals in the design and construction process'. On the presumption that indexicality refers to the interpretation of the listed events referred to in clauses, one can see why Clegg might conclude that it is the contract that causes conflict. However, three points readily arise. First, if the contract does not attempt to list the issues, what will take its place? The listing of contract conditions is both its weakness and its strength. Secondly and more importantly, it is the exploitation by the professionals that causes the problem. Therefore unless exploitation must inevitably occur, one can question whether it is the contract itself or the interpreters of it that are the problem. Finally, the quote from Clegg mentions 'self interested professionals' and perhaps should not be so restrictive, in that many situations arise from the parties' own actions, notwithstanding the advice of their professionals.

Allocation of risk

Likewise, the allocation of risk within the contractual framework should not in theory provide either conflict or dispute. Again in practice, a claim to exercise one's rights under the contract is often perceived as a dispute or can develop into one. A major contention usually arises as a consequence of a risk proving greater than that expected by the party carrying the risk. This again is not necessarily because of the contract itself; however, the inappropriate allocation of risk does seem to create more than its fair share of disputes. Risk and uncertainty do seem to lead to conflict because things seldom turn out as anticipated. Some of the risk-shifting provisions in contracts have fostered many disputes and it is maintained[5] that attempts to shift risk to those who have little or no control over events have been responsible for the proliferation of disputes in the American construction industry. Perhaps fewer disputes would emerge if proper risk analysis was undertaken within an appropriate tendering procedure, and this in turn would alleviate conflict.

The allocation of risk is always a major issue addressed by the contracting parties. The unfair allocation of risks on a construction project will promote later disputes but the authors believe that if the parties are of equal bargaining power there is nothing inherently unfair with any determined allocation of risk, so long as the parties are fully aware of the facts. It may not be unfair but it may nevertheless be inappropriate and a clear distinction should be made. The issue of unfairness becomes much more problematical where the parties are not of equal bargaining power and one might argue that in an industry such as construction, where there frequently exists an excess of supply over demand, this seldom occurs. Certainly, Latham[6] is of the view that in practice, market forces usually make one party dominant. It is in these circumstances that an unfair and inappropriate allocation of risk can result.

Any apportionment of risk leads to a potential problem in that it leads to a dividing line being drawn and as such lines can seldom be drawn in absolute terms, they are often the very place at which disputes and indeed conflict can emerge. It can be argued that in this sense it is far better to place the whole burden of risk on one party. This is one of the arguments that originally inspired design and build with the concept of unitary or single point responsibility.

On the one hand, requiring one party to accept the whole risk reduces the potential for dispute, but on the other hand the unfair allocation of risk often leads to disputes and indeed conflict. The best way forward is generally for the parties best able to control and/or regulate the consequences of the risk, to carry the risk. There may emerge, however, a separation between the ability to evaluate potential risk and the control of the occurrence of the risk. Furthermore, there is the added difficulty that this risk can fall on the party least able to sustain the loss that may result.

It is preferable that a contract should expressly define the allocation of risk, and the consequences of encountering the risk should be properly thought through by the parties. The potential for disputes and or conflicts is then generally reduced. If the allocation of risk is not clearly defined, because of lack of clarity in intent, poor drafting or leaving the risk allocation to be implied, then clearly the potential for problems increases. As stated by Allen[7], 'A key ingredient of any dispute is when one party's expectation differs from another's intention'.

However, even where a contract clearly defines the risk allocation, claims, disputes and conflict can still arise. Claims may occur simply because that is what the contract is designed to do. Alternatively, it may be because the consequences flowing from the occurrence of a risk are not 'consciously' anticipated and the party allocated the risk is seeking to shift position, claiming that the consequences were not reasonably foreseeable at the time of entering into the contract. This situation is aggravated significantly by a tendering procedure that creates a one-off relationship and which does not seek to build a long-term association. Furthermore, a tender won on price alone and in competition for a one-off project, will increase further the potential for claims, simply in order to redress the profitability of the organisation. Theoretically this should not occur, because tenders should be realistic and contracting parties should not seek redress for their own shortcomings, but that of course is an idealistic view. Obviously, anyone whose business is threatened will endeavour to put right the situation, especially when they see it as having been unfairly created through unfair risk allocation, unrealistic levels of competition or short term business relationships.

Standard forms of contract and their drafting

Contracts should reflect the parties' requirements and this is achieved in a variety of ways. The coverage and drafting of contracts varies significantly and this can create confusion. Where contracts seek to deal with particular aspects of a procurement route the difference is acceptable, but where different terminology is being used to achieve similar concepts, it is indefensible.

The existence of a plethora of standard forms of contract, many of which, even within the same suite, use different terminology for the same basic concept raises the question of our whole approach to contract. There is a belief[8] that the number of standard forms and amendments to them generates potential for dispute. This point is emphasised by Baden Hellard[9] when he says that, 'There are now 94 different "standard" contract forms covering the customs and practices which have developed within the industry. These reflect the commercial conditions prevailing within the industry more than they do the legal environment outside the industry.' This latter point was also identified in the INCA[10] research project and shows the importance of rationalising industry practices. This may be the fundamental issue that needs addressing before any significant change can be observed in the number of disputes and instances of conflict.

Latham recognises the problem of conditions of contracts but states:

'endlessly refining existing conditions of contract will not solve adversarial problems. A set of basic principles is required on which modern contracts can be based. A complete family of interlocking documents is also required. The New Engineering Contract (NEC) fulfils many of these principles and requirements...'[11]

That means we need a new approach to contracts but more radically, it could suggest a new approach to tendering and contracting in general. It is necessary to change attitudes as well as the way things are currently done. This implies, among other things, a point referred to earlier, that the conditions must accurately reflect requirements. Whether one standard form of contract, or a small number of forms such as the NEC uses, and based on core clauses will achieve these things is debatable. Nevertheless, it would seem the NEC is at least moving in the right direction.

If a nationally approved database of contract clauses were to be created, this could provide the means to ensure the benefits of standardisation, whilst at the same time enabling free market bargaining to be maintained. This would provide the potential for creating 'bespoke' contracts, based on standard terminology, which met the exact needs of the parties insofar as the contract used different clauses and hence different allocations of risk. This approach is not so much seeking to achieve a watertight contract (probably no such thing exists) but rather to better inform the participants and to make it more difficult for a party to renegue, on account of the better relationship formed.

We referred earlier to the fact that many shortcomings of existing contracts could be overcome by the way a contract is viewed and the drafting of the framework to promote solutions rather than the assertion of rights. Whatever the philosophical approach, it is essential that the clauses are clearly expressed to avoid ambiguity, because ambiguity will, more than most things, nurture a dispute even if it does not lead to one.

Standard forms of contract and dispute resolution

Many existing standard forms of contract make provision for the administrator, whether that is the architect, engineer or some other, to act on behalf
of the promoter and also to adjudicate between the parties over a range of
issues. This raises the problem as to whether someone with a dual role of this
kind can act impartially. It is instructive to note the following comments:

> 'When called upon to make a decision between differing contract lan-
> guage interpretations there is a natural tendency to favour that inter-
> pretation which avoids cost to the client-owner . . . Contractors have come
> to expect an engineer's decision to be adverse even when an engineer is
> obligated to make impartial decisions'[12].

This comment will strike a chord and it is not altogether surprising bearing in
mind the case of *Sutcliffe* v. *Thackrah* (1974). This case established that, under
what is generally termed a traditional or conventional contract, the architect is
the employer's agent notwithstanding that he is required to act fairly. Without
doubt, this places an individual in an almost impossible position, especially in
the extreme commercial environment that is frequently experienced. The
person deciding can also find themselves in the position of deliberating on
their own mistakes. Such a situation must inevitably encourage disputes and
worse still nurture an atmosphere of conflict.

The dual role of the architect or engineer under the various forms of
contract virtually negates the significant potential benefit of any clause
under which they may be provided with some form of conciliatory function.
This lack of an effective pre-emptive remedy often means disputes get out of
hand, views become entrenched and the process becomes protracted. The
notion of pre-emptive remedies is important but more than one such remedy
is required because, as we have observed, disputes vary and often demand
different approaches. All contracts should provide for pre-emptive
remedies, as this is one way of speedily attacking the problem before it gets
out of hand, before views become entrenched and before costs escalate
beyond the bounds of reasonableness. Even where contracts do provide for
such remedies care must be taken to ensure they are both operable and
effective. Little point is served if such a mechanism fails to secure the full
support of the parties and those implementing it on their behalf. This means,
among other things, that such provisions should not be easily avoided.

The procurement process and its management

The NEDO report *The Professions in the Construction Industry* stated:

> 'in both industries (building and civil engineering) there was agreement
> that quality is produced when the designer is allowed the time to

continue his search for the best solution until satisfied that he has the right answer'.

This traditionally meant that the designer attempted to complete the design before commencement of construction. In reality, the designer failed to do this because the process of design is dynamic in nature and because time was often so pressing that the construction was started before a final design had been produced. However, the contract and the contract documentation often ignored this fact, leading to inevitable problems.

Alternative approaches to procurement have been sought and sequential design and construction has been substantially replaced by their simultaneous execution. This, together with shifts in perception as to who should carry the risks of construction, plus the changing attitudes towards the professions and the greater professionalism of many contracting organisations, have all created an almost infinite number of approaches. Although these approaches can be grouped under generic headings, such as traditional, design and build, management contracting, develop and construct, the actual process of every construction project is discrete. It is possible that the application of a generic procurement approach to a particular project does in itself cause problems, but the reality is that most contracts are actually a composite of approaches[13]. Therefore, a generic title could be tantamount to misrepresentation and could inevitably lead to subsequent problems and the need for dispute resolution.

There is also a view[14] that changing roles makes it difficult to establish points of reference, and also that the interdependence of players within the construction industry is fragile. Consequently, the potential for disputes and conflict is appreciable. Although an improvement in construction management is evident, the scale of these changes and the change in business ethos have put severe strain on the participants.

Managerial shortcomings are still an integral part of the industry and these failings are exacerbated by the worsening economic climate, certainly in the short term, for UK construction companies. Failings in the management of the construction process have been noted for many years and it is not intended here to review the extensive works which catalogue the various shortcomings of management. Nevertheless, it is worth noting that bad management will generate circumstances that lead to conflict, as the management attempt to rectify the situation for which they are responsible. Good management not only avoids or at least reduces such situations, but where they do arise it ameliorates the position and reduces the potential for conflict.

The responsibility of management operates at a variety of levels but the two major levels are those within the industry and the construction project itself. Good management of the project provides some advantage but clearly good management within the industry is a prerequisite to the complete fulfilment of objectives.

The industry has consistently devised new procurement approaches as a response to manifest and emerging problems but has often spent more time marketing these approaches than applying itself to the intrinsic imperfections.

Any review of the construction industry and the players within it would show that it is fragmented and permits a wide range of relationships. On the one hand this has merit in that the industry is flexible enough to respond to most circumstances, but on the other hand it means that the units do not necessarily work together regularly. This means that efficiency is reduced as is effectiveness, unless time and effort are expended in creating compatible teams. Teamwork is crucial but to recreate it on each contract is in itself inefficient. Any tendering process that aggravates this is questionable from this viewpoint, but the maintenance of teams used to working with each other often means a reduction in the competitive edge sought by many.

Is conflict inevitable?

From the foregoing, it can be seen that the exercise of one's rights under a contract can lead to a dispute and then to conflict. Disputes need not follow claims, nor does conflict need to follow a dispute. Disputes can sometimes, but not invariably, be avoided by skilled personnel who take control of the situation. A genuine dispute may nevertheless emerge and require another means of resolution; if that is what exists it should not be avoided. The only question then is of the most appropriate forum for resolving the dispute. Traditionally arbitration and litigation have been used but increasingly an alternative dispute resolution forum is sought.

Conflict on the other hand can nearly always be avoided and should be avoided. The problem, however, is that it is about people rather than the contract itself. Conflict may manifest itself because of the nature of the parties and their advisers and indeed the chemistry between them. The contract may or may not aggravate this process. The answer to this problem is much more about prevention than about seeking to overcome a conflict that has already arisen. Arbitration and litigation may resolve a dispute but will seldom, if ever, resolve a conflict. Alternative dispute resolution techniques can also resolve disputes and, although this is the primary purpose, their less adversarial approach can also alleviate conflict.

As already mentioned, conflict is created by people, and furthermore disputes are created by people. The former is frequently destructive, leading to a breakdown in relationships, although some argue that conflict is an essential ingredient in the promotion of efficiency. Disputes, however, cannot lead to a breakdown in relationships while they remain as disputes and no more.

The issue of construction conflict management and resolution was addressed in a major way at the First International Construction Management Conference held at UMIST in September 1992. Leading practitioners, academics and others came together to explore a number of problems and in doing so many raised the question as to whether conflict in construction is inevitable. Most of the opinion expressed in the UMIST conference papers is that it is inevitable[15]. There is adequate evidence to support this view.

Smith[16] is of the view that '... construction conflict is not ... to be regarded solely as a one off situation concerning two parties in isolation. Construction conflicts are, after all, endemic in the industry. The reasons for them flow from the way the industry functions...'. Baden Hellard[17] agrees that problems exist within the industry when he says that 'The organisation of the construction industry today has a built-in recipe for conflict'.

Langford *et al.*[18] also believe that conflict between contracting companies may be inevitable. This raises another dimension to the problem in that conflict is often perceived as between contractor and promoter (and or the promoter's consultants) but conflict is equally, if not more so, virulent among contractors and subcontractors. Certainly, Harding[19] established that conflict between these parties was common.

Many studies[20], have been undertaken into why claims occur and it is apparent that there is no one cause. Causes range from gaps in information to a deliberate change of mind, from ground conditions to unforeseeable events. It is also clear that one can see in a claim the potential for dispute and conflict. If claims can be avoided, disputes and conflict will diminish.

The number of cases going before the London Official Referees' Courts[21] between 1973 and 1990 grew substantially and if this were taken as an indicator, one might conclude that not only is 'conflict' in the form of disputes inevitable but that it is increasing. However, it is not possible to deduce this from such data. Cases before the courts represent only a small percentage of all the disputes that arise and therefore without other supporting evidence one can only conclude that the courts are hearing more cases than they did. This may be because the courts are becoming more popular as a resolution process and are preferred to the alternatives. It may be because disputants are becoming more litigious or because the disputes are more complex with potentially larger sums at stake. There is little reliable evidence to support a particular hypothesis but this increase has taken place against a growing realisation that pursuing a dispute in court is both extremely time consuming and costly. Does this suggest that the finality of this process outweighs the purported benefits of the alternatives? Does it suggest that justice, whatever that may be, is preferable to a quick-fix compromise solution?

It is evident from the available data and opinions that conflict is inevitable in any contract. Competition will exist in some form in every contract, regardless of the procurement path and tendering process. The degree of competition will vary appreciably and one might speculate that the less the competition is explicit, the fewer disputes will emerge and the fewer of these will involve conflict. Underlying this is whether an anticipated higher initial cost is preferable to the less certain final cost that higher levels of competition may produce. The difficulty in determining final cost where competition is high is that the realisation of a problem is virtually indeterminable, as is the amount of resource to be applied to its resolution should the event occur.

Using and alleviating conflict

Although conflict is inevitable, it is not necessarily something that should be viewed as solely disadvantageous. It may be that a certain level of conflict in an organisation is not only inevitable but also desirable, for conflict is both a cause and effect of change. It has also been argued[22] that functional conflict is an inescapable part of the contract system we have chosen and so one may conclude that it is necessary. Whether it is desirable would be another matter, but clearly it is inherent in any interaction that is required to achieve objectives, whether these objectives are common or not. Smith[23] distinguishes between functional conflict and dysfunctional conflict and summarises the difference as:

'Functional conflict is essentially a construction community problem, when it is an inescapable consequence of our trading relationships. Dysfunctional conflict may have arisen if the actions of the parties have gone beyond what we may recognise as a functional conflict.'

It is not easy to isolate functional conflict in that one has to be able to provide an industry benchmark for what is acceptable practice and behaviour. The multifarious approach to procurement makes this difficult, as does the shifting moral and ethical base upon which our actions are founded. Nevertheless, some rationalisation of our current systems would assist in arriving at an agreed position. However, whatever progress is made here it would seem that our contracts (in the legal sense) do lend themselves to functional conflict, notwithstanding the fact that theoretically they are established by agreement. The subsequent use to which they are put creates the problem and frequently moves us to a dysfunctional position. There is a thin line between the two and at present there is not always consensus as to when it has been crossed. This raises the subsequent problem of identifying the best means to resolve the problem that has emerged.

Rahim[24] seems to accept organisational conflict and believes that it need not necessarily be reduced, suppressed or eliminated. He believes that instead one should attempt to manage the process. In order to manage the process one needs to understand what conflict is, how it occurs, and how it is resolved. To provide this knowledge, understanding and the appropriate skills we need educational support, and the current training of construction professionals and others does not provide this. Hence, it would seem that until this particular point is addressed we have a situation where inadequately prepared persons will attempt to resolve inevitable conflicts.

As we have noted, the dual role of some consultants often works against the possibility of arriving at a solution using the conciliatory procedures that may be provided. This is one area that could be remedied, either by changing the functions of the consultants or by adopting procurement processes which distinguish between the role of agent and adjudicator.

It can be seen that conflict in construction contracts can be scrutinised from the perspective of preventing and also from that of controlling. It has been acknowledged that conflict is latent in any contractual arrangement but also that the potential for claims and disputes and hence conflict may be substantially reduced. The opportunity for reducing the potential can occur at two distinct stages: pre-contract and post-contract. The above discussion indicates that a reduction in the potential for conflict at pre-contract can be achieved by appropriate decisions with regard to:

- allocation of risk
- tendering process
- selection of the team charged with responsibility for delivering the project
- documentation (including the formation of the contract and especially the inclusion of pre-emptive remedies)
- communication.

This is not an exhaustive list of preventive measures but more an indicative list of the important ones. Although this chapter is primarily about dealing with the problems of claims, disputes and conflict arising at the post-contract stage, one cannot dismiss the pre-contract stage without exploring a little further matters relating to pre-contract prevention.

The decisions made in respect of the five points above set the economic and social environment for the contract.

Construction contracts have inherent risks and the appropriate apportionment of these risks is a major challenge, especially as there is disagreement as to what constitutes fairness. This is exacerbated by economic conditions at the time of entering the contract and by those prevailing during the performance of the contract, the latter being especially problematic when they differ from those anticipated and accounted for at the tender stage. To improve the situation a number of issues need to be acknowledged and the following have a bearing on the extent of potential for conflict:

- the means to allocate risks appropriately
- the means to accurately communicate the nature and boundaries of risks
- the means to assess risks
- the opportunity to price fairly the risks
- the means to control the consequences of risks should they arise
- the acceptance of the agreement regarding these risk
- the ability to carry the financial consequences of the occurrence of a risk.

The tendering process itself will affect risk carried by the parties. Risk in any tendering process is, to some degree, dependent on the extent of competition that is determined by market conditions. As previously discussed, too much competition will aggravate the potential for conflict, whereas too little competition has its own problems and indeed may create demands from the client that will also lead to conflict.

Any tendering process that keeps the parties at arm's length is more likely to suffer from disputes. This means that the involvement of both parties is desirable and negotiation prior to contract has much to offer. The scale of negotiation ranges from full negotiation with one nominated contractor to negotiation of outstanding details following open competitive tendering. The former provides greater opportunity for co-operation and perhaps more importantly it creates the ambience to develop understanding relationships. The involvement of parties in this type of situation aids what is often described as 'bonding'.

The selection of the project team also has an important bearing on this last point. There are two major relationships concerning the project team: the intra-relationship and the inter-relationship with the construction promoter.

Clearly, if the chemistry is right between all the participants, then the potential for conflict is reduced. Where teams have worked together the result of the chemistry is known, albeit not in absolute terms because relationships can and do change. Nevertheless, this is often better than the risk of putting together participants unknown to each other. It is possible to use psychological techniques to improve the selection process but at present their use in construction is limited and their success unproven. The problem with construction work is that by its nature different teams frequently emerge for each project, even when the same practices are involved. Where teams are held together there is the real chance to form 'bonding' between the participants and this can lead to distinct benefits both for the promoter and the consultants. The benefit to the promoter flows from the improved communication and relationships that in turn reduce the incidence of disputes. There is a potential downside to the use of the same project team in that they can work for each other more than they do for the promoter.

'Bonding' and 'partnering' are perhaps even more important between the promoter and the project team, which includes the constructor. If such development reduces conflict it is not difficult to understand why 'partnering' is becoming a topic of earnest discussion. An important aspect of 'bonding' and 'partnering' is that they both require time and effort to establish, something which many feel is a luxury in a competitive environment.

This last point raises the fundamental philosophical issue of how we procure buildings. The real benefits of this form of 'bonding' and 'partnering' can only generally be seen where there is a continuous building programme. Where such a programme does not exist and ad hoc building works are involved one recognises the disinclination to spend time and effort, when it is perceived that it will not usually provide real benefit within the time frame of one project. This is especially true with consumers involved with small-scale projects, but these represent a significant number of projects. Such a perception may be flawed in that any cost benefit analysis tends to assume that the contract will run without problems. Evidence suggests that this is unlikely to be the case unless one invests in creating the right project environment. The difficulty for the one-off client is that the arithmetic of this problem is not straightforward and consequently there is a tendency to act, albeit often subconsciously, as one's own 'insurer'. However, when a problem emerges the natural reaction of the client, as with any insurer, is to resist the consequences, thus creating circumstances which are conducive to disputes and conflict.

Project documentation is an essential requirement and the way this is compiled and communicated is generally a prerequisite to a successful outcome. It is acknowledged that some successful projects, in the eyes of the parties, have little or no documentation. Despite this, one cannot recommend such an approach. In order to reduce the potential for dispute and conflict one must ensure the following:

- an adequate brief is provided by the promoter
- adequate time is given for the preparation of tender documentation
- the documents effectively communicate the promoter's requirements.

Good communication between all the participants is essential. The earlier this starts the better the chances of overcoming problems of definition and interpretation.

Pre-contract prevention is a vitally important part of the process. However, even the most effective pre-contract prevention is unlikely to succeed totally in avoiding the use of post contract remedies and procedures. Consequently, every contract should address the issue of how the parties will resolve differences should they occur. With the advent of the Housing Grants, Construction and Regeneration Act 1996 (HGCRA), adjudication provisions are statutory on a wide range of construction contracts (see Chapter 9) but one still needs to consider whether to provide for arbitration (frequently less efficient and more costly than litigation) and whether to incorporate pre-emptive remedies such as mediation or conciliation. The use of such remedies immediately a difference occurs significantly reduces the prospect of the difference escalating into a greater problem, but how this is provided is made harder because of the HGCRA.

Chapter 2
Arbitration and Litigation:
A Tarnished Reputation

Problems of traditional dispute resolution

A former Chief Justice of the USA, Warren Burger, once said:

> 'The obligation of our profession is ... to serve as healers of human conflict. To fulfil our traditional obligation means that we should provide mechanisms that can produce an acceptable result in the shortest possible time, with the least possible expense and with the minimum of stress on the participants. That is what justice is all about.'[1]

Leaving to one side the possibly conflicting emotions of those involved in the dispute resolution business, where certain lawyers need to question the value of a 'litigation client who is not litigating', no sensible person could dissent from Warren Burger's comments. Dispute resolution is a service industry and must recognise client needs. This theme has been taken up by many leading members of the judiciary and was the cornerstone of Lord Woolf's interim and final reviews of civil litigation[2], *Access to Justice*, and is reflected in the new civil litigation rules, the Civil Procedure Rules (CPR), operative from 26 April 1999. Another very senior judge, the former Lord Chief Justice, Lord Taylor, rightly said in identifying the obligation of the legal profession to be responsible in plying its trade:

> 'A trial is not a game. The role of a judge should not be restricted to that of an umpire sitting well above the play, intervening only to restrain intemperate language and racket throwing.'[3]

Lord Taylor's remark echoes earlier words of Warren Burger who on one occasion said:

> 'Trials by the adversarial contest must in time go the way of ancient trial by battle and blood.'[4]

Increasingly, litigation, the legal process and the ways of lawyers and expert witnesses have attracted a very poor press. Particularly in the USA lawyers are the butt of savage jokes which emphasise their reputed avarice and self-interest. Even in the UK a survey of the top 400 companies in *The Times Top*

1000, asking for their views on litigation, found widespread criticism of the length, complexity and cost of civil justice:

> 'A substantial majority (70%) suggested the whole system takes too long, whilst almost 40% suggested that the costs of litigation are far too high.'[5]

Companies apparently support a shift away from oral advocacy in civil trials and a greater emphasis on written submissions. They want judges, not the parties to the dispute, to control the pace of proceedings and to determine how long cases should take. This accords with case management under Civil Procedure Rules. More than 60% of the firms questioned favoured a paper trial instead of one based on oral argument and evidence; and a substantial majority wanted the control of the timetable of cases given to a *procedural* judge.

Over the years the construction industry trade press has frequently reported criticism of the legal process. For instance, Ian Dixon, chairman of Wilmott Dixon and the then chairman of the Construction Industry Council, was quoted on one occasion as saying:

> 'You can't win if you go to court. The high legal costs are part of it. Litigation is long and repetitive. The legal system is abysmal and inefficient.'[6]

Prior to the HGCRA with the notable exceptions of the New Engineering Contract, 2nd Edition, with its reliance on adjudication, and the IChemE with expert determination as the initial form of dispute resolution, and to a lesser extent the ICE contracts which, for a number of years, had provided for conciliation, the principal dispute resolution method traditionally anticipated by most other standard form construction contracts was arbitration. The emphasis may well change, however, following the statutory right to adjudication introduced by the Housing Grants, Construction and Regeneration Act 1996 for a wide range of UK construction contracts. The demise of *Derek Crouch* may also assist the decline of arbitration. Arbitration, traditionally seen as a 'no nonsense' method of dispute resolution, relying more on technical assessment than on the application of judicial nuances, has for a number of years been regarded generally as overly legalistic. In *Northern Regional Health Authority* v. *Derek Crouch Construction Company Limited* (1984), overturned by the House of Lords in 1998 on other grounds, Sir John Donaldson MR stated:

> 'Arbitration is usually no more and no less than litigation in the private sector.'[7]

The conclusion reached in an Australian research report[8] into claims and disputes in the construction industry, was that:

> '...arbitration has broken down as a cheap and efficient means of resolving construction disputes, albeit that the cause may be the strenuously

adversarial manner in which the disputants themselves pursue the arbitral process.'

Perhaps the apparent inability to resolve disputes efficiently and cost
effectively stems from the way in which we train our lawyers and the
alacrity with which many expert witnesses become part of the legal process.
Ours may be an adversarial system of law, but the words of a president of
Columbia University are significant:

> 'The idea that we should spend all our time in law school teaching people
> how to win instead of how to settle is very damaging'.[9]

It is however important to strike a balance in any assessment of litigation and
arbitration and to resist the temptation to conclude that they are never successful or in a client's best interests. On occasions, a claimant is pursuing
outstanding monies from an unprincipled opponent who has no intention of
negotiating sensibly. Frequently facts are complex and can justifiably produce
different but ostensibly valid interpretations. Less often cases throw up points
of contract interpretation, and an analysis of legal principles is necessary.
Litigation, including summary judgment under RSC Order 14 (now CPR Part
24) and interim payments under RSC Order 29 (now CPR Part 25), although
now rendered less effective in the construction industry by section 9 of the
Arbitration Act 1996, can be an extremely effective means of debt collecting
where the issues are factually or legally clear-cut. Unfortunately this is rarely
the case in commercial disputes, a fact well known to the parties (at least if
they are honest with themselves) and one which some lawyers appear to
ignore for as long as possible. It is sometimes claimed that litigation or
arbitration provide a lever on the defendant to make him negotiate and
enhance the prospects of a quicker negotiated settlement. Such a tactic often
fails. The parties become more and more entrenched in the litigation until the
dispute is resolved by way of a belated piece of 'horse-trading'. This happens
only when high legal costs have been incurred, vast amounts of time have
been expended by professional advisers and client on case preparation and
the parties' litigation enthusiasm has been completely sapped.
  All too often the effects of litigation and arbitration are:

- polarised positions
- a drain on the client's managerial time
- clients who feel out of touch with their own dispute and the victims of a
  legal take-over
- damaged commercial relationships
- expensive and long drawn-out proceedings
- use of deliberate delaying tactics by a defendant who knows how to play
  the system
- a pyrrhic victory for the successful litigant with monies recovered representing a mere fraction of actual expenditure

- a judgment that is impossible to enforce because there is not the means to satisfy it
- a belated realisation by the plaintiff that the principal reason for the litigation or arbitration was the impecuniosity of the defendant
- lawyers who are reluctant to engage in early 'reality testing' with their client

The perceived failure of litigation and arbitration, first in the USA and then in other jurisdictions, has encouraged the rise of ADR. There are three main types of ADR: mediation, conciliation and mini-trial. These are briefly outlined below and are dealt with at length in Chapters 4 and 7.

Mediation

An independent third party, the mediator, assists the parties through individual meetings with them (caucuses) as well as joint sessions (a form of shuttle diplomacy) to focus on their *real* interests and strengths as opposed to their emotions in an attempt to draw them together towards possible settlement. Crucial to the mediation process is that the independent third party ordinarily does not make recommendations as to what would be an appropriate settlement. He is merely there to assist the parties to find and settle their own agreement. The mediator is quite different from an adjudicator who is called upon to make a decision.

Conciliation

A conciliator is usually more interventionist than a mediator but still endeavours to bring disputing parties together and assist them to focus on the key issues. Conciliation has been well known in the UK in employment matters for a number of years via ACAS. Given the looseness of ADR terminology the terms 'mediator' and 'conciliator' are often used interchangeably.

Mini-trial (executive tribunal)

Each party presents the issues to senior executives of the disputing parties, who are often assisted by a neutral chairman. Lawyers may but not necessarily represent the parties. The chairman, perhaps a lawyer, may advise on the likely outcome of litigation without any binding authority on the parties. After presentation of the issues, the executives try to negotiate a settlement. If successful, the settlement is often set out in a legally enforceable written document. The mini-trial is a misnomer to the extent that it is not really a trial at all. With the legal rules of evidence usually discarded, it is a settlement procedure designed to convert a legal dispute back into a business problem.

Since the late 1980s and early 1990s ADR has been promoted in the UK by a number of bodies, including the Centre for Dispute Resolution (CEDR), with limited success. In 1995 ADR was used to resolve 5% of construction disputes, arbitration 38%, litigation 43%, negotiation 13% and adjudication 1%[10]. Even later, Oxford Brookes University data[11] only shows a 4% take up for ADR. Such a survey sits uneasily with criticism of litigation and arbitration by construction professionals. Although perhaps disillusioned with litigation and arbitration, a sea change will be required in the culture of the construction industry before ADR is seen as more than 'fringe medicine'. The sea change in respect to adjudication occurred with the HGCRA.

Perhaps the construction industry is not yet ready for principled negotiations. Unfortunately the presentation of many contractor's claims is something of a black art with contractors willingly adopting the clever, often logically superficial, formulations of their own claims consultants. On the other side, many employers do not wish to pay or have no money with which to pay. Arbitration or litigation then become equally attractive to both contractor and employer. The contractor believes that he will be able to persuade a court or arbitrator of the legitimacy of his claim by a clever presentation, while at the same time certain employers view the slow and expensive legal process as a means of denying a contractor his legitimate and reasonable financial expectations. Employer and contractor may both have entered into a particular contract following an untenable financial analysis. The employer has underestimated to a funding institution the real cost of the project (so as not to jeopardise his chances of securing funding) while the contractor, equally desperate, has adopted negative margins with the intention of making up the difference later by claims.

Construction professionals must go back to basics and question the adequacy of the present procurement systems, even allowing for the various uncertainties that construction projects inevitably create. Is the traditional approach to procurement by the professionals now outmoded? Can ad hoc alliances of consultants, contractors and subcontractors honestly form a harmonious team when everyone knows that lowest price secured the job in the first place. Certainly it would appear that continental Europe, which adopts more of a construction management type approach and where lowest tender does not automatically win, creates fewer claims, although the same cannot be said of the USA which has been the prime promoter of the construction management system. At the risk of perhaps resorting to a stereotype it may be that English and American contractors are simply more litigious than their continental counterparts. In truth, the problems of the construction industry must be more complex than that and difficulties that arise during the course of contracts must have their origins in the whole procurement method, including the tendering process. Governmental awareness that all is not well with the world was reflected in the setting up of the inquiry under Sir Michael Latham, which led initially to the provisional report, *Trust and Money*[12], and subsequently his full report, *Constructing the*

Team[13]. Now it really is the time for solutions, although was not the same said after the *Banwell Report*[14] in 1964?

The challenge of ADR

In High Court litigation lawyers have for some years been unable to ignore the need at least to pay lip service to ADR. The most important change prior to implementation of any of the reforms proposed by Lord Woolf and embodied in the Civil Procedure Rules (CPR) operative after 26 April 1999, the civil justice revolution, was the *Practice Note (Civil Litigation: Case Management)* [1995] 1 AER 385 which was announced by Lord Taylor of Gosforth, the Lord Chief Justice, and Sir Richard Scott, the Vice-Chancellor, on 24 January 1995 for use in both the Queen's Bench (including by implication the Official Referee's Court) and Chancery Divisions of the High Court. It contained a number of proposals, the aim of which is to streamline civil litigation procedures in the High Court and created a greater move towards case management summarised in the following policy statement:

> 'The paramount importance of reducing the cost and delay of civil litigation makes it necessary for judges sitting at first instance to assert greater control over the preparation for and conduct of hearings than has hitherto been customary. Important for litigation lawyers is the threat that a failure to conduct cases economically will lead to orders for costs, including a wasted costs order, against such practitioners.'

The Practice Note repeats the earlier *Practice Statement (Commercial Court: Alternative Dispute Resolution)* [1994] 1 WLR 14, which recognised the value of ADR. ADR is referred to in the Practice Note under paragraphs 10, 11 and 12:

- Have you or Counsel discussed with your client/clients the possibility of attempting to resolve the dispute by Alternative Dispute Resolution?
- Would some form of Alternative Dispute Resolution resolve or narrow the issues?
- Have you or your client/clients explored with the other parties the possibility of resolving the dispute by Alternative Dispute Resolution?

Where the Practice Note is weak is in not referring to possible sanctions for non-compliance. The cynical solicitor could neuter the application of the ADR provisions either by refusing to discuss them with a client or by always setting the benchmark that his or her cases were ill-suited to ADR. Even among the specialist construction law judges, the Official Referees, now judges of the Technology and Construction Court, the support for ADR appears patchy. Only the now retired Official Referee, Judge Fox-Andrews QC, included a direction in his orders on a summons for directions informing the parties of the value of considering resolution of their disputes via one of the ADR techniques.

On 7 June 1996, the Commercial Court under Waller J issued what was, in English terms, a very strongly worded endorsement of ADR in its *Practice Statement (Commercial Cases: Alternative Dispute Resolution)(No 2)* [1996] 1 WLR 1024. As the Practice Statement indicates:

> 'The Judges of the Commercial Court, in conjunction with the Commercial Court Committee, have recently considered whether it is now desirable that any further steps should be taken to encourage the wider use of ADR as a means of settling disputes pending before the Court.'

The Commercial Court identifies five factors, which may encourage the use of ADR. These are:

(1)   a significant reduction in cost
(2)   a reduction in delays in achieving finality
(3)   the preservation of existing commercial relationships and market reputation
(4)   a greater range of settlement solutions than those offered by litigation
(5)   a substantial contribution to the more efficient use of judicial resources.

However, what is different from any previous judicial comment is the assertion in the Practice Statement that judges of the Commercial Court will positively encourage parties to adopt ADR:

> 'If it should appear to the Judge that the action before him or any of the issues arising in it are particularly appropriate for an attempt at settlement by ADR techniques but that the parties have not previously attempted settlement by such means, he may invite the parties to take positive steps to set in motion ADR procedures. The Judge may, if he considers it appropriate, adjourn the proceedings then before him for a specified period of time to encourage and enable the parties to take such steps. He may, for this purpose, extend the time for compliance by the parties, or either of them, with any requirement of the Rules of the Supreme Court, or previous interlocutory orders in the proceedings.'

A further radical departure in the Practice Statement is the endorsement of the principle of 'early neutral evaluation'. This is a recognition that judges, particularly under the English adversarial system, have to listen to considerable amounts of evidence which ordinarily it is the parties' right to lead, even if the judge is privately of the opinion that much evidence will not greatly assist him in coming to an appropriate conclusion. By permitting early neutral evaluation, lengthy trials may be curtailed. Under the Practice Statement, the assigned judge of the Commercial Court may provide the evaluation, or arrange for another judge to do so. Of course, the Judge cannot impose early neutral evaluation on the parties unless the parties otherwise agree. If there is early neutral evaluation, which does not result in settlement, the particular judge will not take any further part in the proceedings.

Where the Practice Statement is also tougher is the wide discretion given to judges assessing costs. The judge may consider that ADR has been deployed. The parties must report back to the Judge if ADR is attempted and fails, although, quite properly, the substantive contact between the parties and their advisers is not to be made known to the judge.

ADR has now been tried out to a limited extent in the County Court[15]. From May 1996 the Central London Court operated a one-year pilot scheme to allow mediation of certain civil disputes. The scheme applied to disputes in the £3000–£10,000 range. Parties who opted for mediation did so without prejudice to their court-based rights and had the benefit of a three-hour session with a trained mediator from one of the ADR providers outside court hours between 4.30PM and 7.30PM. Each side paid £25 towards the mediator's costs. The mediation was arranged within 28 days. A similar but two-year scheme operated in the Patents County Court in London. In addition, the Patents County Court is offering arbitration by a technical arbitrator included in a court list. Also, the Lord Chancellor's Department[16] has produced a useful booklet on resolving disputes by methods other than litigation.

Under the CPR, ADR is given a boost. Rule 26.4(2) and (3) allows the court to stay the whole or part of any proceedings generally or to a specific date, while Rule 26.4(2) allows a party, filing the completed allocation questionnaire, to make a written request for the proceedings to be stayed to allow the use of ADR. If litigation is unnecessarily pursued the court can penalise a party in costs under Rule 44.5(3).

Can arbitration ever be simple?

As already stated earlier in this chapter, traditionally most construction contracts have anticipated that dispute resolution would be by means of arbitration. Over recent years the problems of arbitration have been much debated. Certain construction professionals believe that arbitration has been hijacked by lawyers and turned into something akin to the litigation process in the High Court. Such people remember, or choose to remember, an earlier period when arbitration was a cheap and quick method of dispute resolution, shorn of much of the paraphernalia of the legal process.

The obvious question is, can arbitration ever be really different from litigation? There is a danger that the present clamour for a 'back to basics' approach is, although well intentioned, an exercise in unreality. Essentially, arbitration like litigation, is designed to identify and establish the *legal* rights and obligations of the parties. It is not some form of ADR. Construction disputes are usually complex, not least those relating to loss and expense claims or building defects. Apart from applying the law correctly to a given set of facts, an arbitrator, no less than the judge, needs to examine facts that interrelate and combine, where causation needs to be demonstrated and proved. A common, but frequently difficult, problem that arises in the analysis of construction claims is the effect of concurrent delays. Not all of these will necessarily be the employer's responsibility.

The complexity of construction disputes means that they rely heavily on witnesses of fact appearing before a judge or arbitrator, with their recollections and suppositions being subjected to cross-examination. Too many money claims in the construction industry commence on the basis that a project was programmed to complete in x weeks, only to be completed in y weeks. The contract value was £x, whereas the 'as built' costs were £y. The contractor starts from the glib proposition that he is entitled to the difference between the contract value and the 'as built' cost as his direct loss and/or expense. However much additional documentation is provided by claims consultants, many claims are, beneath the reams of paper, little more than global in their composition. To analyse such claims properly in the adversarial climate which arbitration shares with litigation, is inevitably time consuming and expensive. For this reason parties who commence an arbitration thinking it is more cost effective than litigation may end up with a hearing of several weeks duration or more and a costs bill of several hundred thousand pounds. An example is the claim for £20,000 which went to final hearing before a well-known construction arbitrator, with a total costs bill of £250,000. By contrast, in High Court litigation the judge and the court buildings continue to come free, at least for the moment, although moves to a self-financing courts' service may radically alter this.

Further, construction professionals should not forget that arbitration law and practice have long been part of the legal process. Since 1695 arbitration in England and Wales has been subject to legislative control. In recent years control was found in the Arbitration Acts 1950, 1975 and 1979. The need for a new consolidating statute to replace these three Acts was recognised and a first draft consultative bill was produced after considerable initial soundings[17]. Because of the intense hostility that greeted the departmental advisory committee's first attempt at providing a modern framework for arbitration, a second draft bill was produced in July 1995 together with a consultative paper[18]. The Arbitration Act 1996 received Royal Assent in July 1996 and came into force on 31 January 1997. The new legislation does not remove arbitration from the legal process. The original draft bill was criticised for retaining what were perceived by many as the outmoded and legalistic principles of the present legislation. Comment about the second draft bill[19] and the Act itself was considerably more favourable.

In addition to direct legislative control, arbitration is subject to a plethora of case law decisions. Many of these cases are complex and contradictory in their application. An illustrative example is the doctrine of arbitral misconduct for the purposes of Section 23 of the 1950 Act. The 1996 Act does not use the word misconduct but section 68 refers to serious irregularity and one is left wondering whether a change of word is the real answer. Furthermore, we will in due course see what other problems emerge under the 1996 Act, but already the issue of legislative control has emerged in *Inco Europe Ltd and Others* v. *First Choice Distribution and Others* (1998). In this case the Court of Appeal held that they had jurisdiction to hear an appeal against the grant or refusal of a stay of proceedings in favour of arbitration under section 9 of the Arbitration Act 1996.

Apart from the excellent ICE Arbitration Procedure (England and Wales) (1983) and (1997) and the much less satisfactory JCT Arbitration Rules 1988, which have been superseded by the requirements of JCT Amendment No. 18 and JCT 98, which are considered below, most lawyers and even many claims consultants have in the past tended to prepare for and run arbitration hearings as if such hearings were trials in the High Court and they have not been truly innovative in their approach. The Arbitration Act 1996 provides an opportunity for this to change.

Perhaps surprisingly, many procedural short cuts used in both litigation and arbitration were pioneered by the Official Referees in the High Court. These include:

- exchange of expert witness reports
- meetings of experts to agree figures as figures and to identify those issues in the dispute that can be agreed and not agreed (RSC Order 38 rule 38)
- the exchange of the witness statements of the witnesses of fact (RSC Order 38 rule 2A) to dispense with or otherwise drastically limit examination-in-chief
- restrictions on the time available to counsel to make oral representations to the tribunal or engage in cross-examination, where the interests of justice so demand.

However, whatever the methods adopted to limit the length of hearings, the need to test the evidence of the other party by detailed cross-examination remains. This is usually the most time consuming aspect of any trial or arbitration hearing.

Attempts to speed up the arbitration process often seem to fail. Counsel and expert witnesses, whose business is the conduct of litigation, frequently become unavailable to commence a hearing on a particular date, although in arbitrations they usually have generous advance warning of the time commitments. Once started, whether a case proceeds by way of conventional pleadings, as understood by lawyers, or by way of seemingly fuller statements of case with all supporting documentation included in the statement of case, the lawyers' and claims consultants' desire for further information by way of further and better particulars never appears to be fully satiated. In all litigation, including arbitration, requests for further and better particulars and replies to them, frequently difficult to resist, often delay the process of bringing the matter to trial or final hearing.

One of the greatest defects in English arbitration law has been the dead hand effect of legislative control and possible intervention from the courts provided by the Arbitration Acts 1950–1979. This problem is not satisfactorily addressed in the Arbitration Act 1996, with the courts retaining under sections 42–45 powers in regard to arbitration proceedings and under sections 66–71 in regard to arbitral awards. Interestingly, under section 46 the arbitrator may with the consent of the parties sit as an amiable compositeur. This is a radical departure for the UK although recognised in other jurisdictions.

Arbitration rules

Concern at the adverse reputation arbitration was acquiring in the late 1980s for:

- slowness
- too legalistic an approach
- expense

led the JCT in 1988 to introduce new arbitration rules to allow greater flexibility in arbitrations arising under the JCT contracts. The JCT have now adopted the Construction Industry Model Arbitration Rules (CIMAR) first published in February 1998, but the earlier rules are briefly discussed here because they may remain relevant for ongoing references.

The JCT Arbitration Rules 1988 may still apply if required but this is unlikely as the JCT edition of CIMAR is an amendment to the earlier rules. The 1988 rules envisage a dispute being settled in one of three ways:

(1) written statements only without a hearing (a rule 5 arbitration)
(2) written statements plus a hearing (a rule 6 arbitration)
(3) a short procedure with hearing (a rule 7 arbitration).

It is open to the parties to decide which procedure to follow. If they cannot do so, then the dispute will be settled under rule 5 unless the arbitrator is persuaded that it would be more applicable for the dispute to be settled under rule 6. If the claimant wants a rule 7 arbitration, he must formulate his case in sufficient detail to persuade the respondent to agree to a hearing under this rule. If the respondent is not so persuaded, then the arbitrator must choose between a rule 5 or a rule 6 arbitration. The arbitrator cannot require a rule 7 arbitration. Similarly, a claimant requires the consent of the respondent before there can be a rule 7 arbitration.

Although flexibility was built into the JCT rules, the appropriateness of a 'documents only' arbitration (a rule 5 arbitration) may be limited to situations where the correspondence is self-explanatory, comprehensive and factually non-contentious and matters of contract interpretation are in issue. This would appear to rule out a 'documents only' approach in almost all construction disputes, including loss and expense claims. It is further open to question whether an arbitrator may be criticised for not requiring a hearing under rule 6 when one party only is in favour of a 'documents only' hearing under rule 5. The great advantage of the JCT rules was supposed to be to ensure a speedy outcome to arbitrations. Slowness and cost, as previously stated, are often taken to be the two most negative hallmarks of arbitration. Provided that the relevant rules are complied with, the period in question to conclude an arbitration varies from 112 days for a rule 5 arbitration to 154 days plus the period to and length of the hearing for a rule 6 arbitration and 49 days for a rule 7 arbitration.

The JCT rules promote speed in two ways. First, if a statement is not served by the due date for a rule 5 or a rule 6 arbitration, the arbitrator is required to give to the defaulting party formal notice that he has a further seven days in which to comply and, if he does not, the arbitrator may proceed without him. A claimant who fails to serve a statement of case on time can find the arbitrator dismissing the claim and ordering the claimant to pay the fees and expenses of the arbitrator and any costs incurred by the respondent. Moreover, if the defaulting party's statement is delivered late, the arbitrator is required to disregard it unless the defaulting party can give a good and proper reason why the statement was not served by him on time and why a request for an extension of time was not properly made. In this regard an extension of time will only be given if the arbitrator is satisfied that the need for such an extension arose on account of a matter which could reasonably be considered to be outside the control of the party concerned. The rules, which require service of the various pleadings within 14 days for a rule 5 arbitration and 28 days for a rule 6 arbitration, are somewhat loaded against an employer who receives a loss and expense claim from a contractor many months after the works are completed. In truth, the JCT rules time periods are too short to investigate claims and formulate responses in detail, particularly where an employer's in-house personnel and outside advisers need to be interviewed. The advice of many lawyers to employers is to discard the JCT rules completely.

The second way in which the JCT rules promote speed is that the pleadings of the party should be:

- comprehensive documents setting out the nature of the case being made
- the facts relied on
- the legal basis of the claim or counterclaim.

The JCT rules further require a party serving a particular pleading to list those documents which are considered necessary to support its case (in the case of a rule 5 or rule 6 arbitration) and supply a copy of the relevant documents with the necessary passages highlighted. The JCT rules are also designed to reduce discovery and inspection (again an expensive process in traditional arbitration and litigation), although the arbitrator has a residual power (rule 12.1.7) to require a party to give further and better discovery.

The Construction Industry Model Arbitration (CIMAR) rules, which the JCT has now adopted, were produced specifically for use with arbitration agreements within the construction industry under the Arbitration Act 1996. These rules contain much of what is contained within the Act and also aim to:

(1)   extend or amend the provisions of the Act where necessary
(2)   add a general framework to the specific powers and duties in order to provide guidance to users as well as to arbitrators.[20]

The approach adopted is supposed to couple brevity with clarity and to promote user friendliness. Although the CIMAR rules are quite different from the JCT Arbitration Rules 1988, they maintain the use of one of the three procedures and build on the concept of flexibility, which of course is what the Arbitration Act 1996 is seeking to achieve.

The three procedures for conducting arbitration under the CIMAR rules differ from the old JCT rules. The form of procedure is dealt with under rule 6. Under clause 6.1 the arbitrator must consider which is the most appropriate procedure. Rule 6.3 requires the arbitrator to convene a meeting with the parties and give directions as to the procedure to be followed, with the rule specifically providing for:

(1)   adoption of the procedures in rules 7, 8 or 9
(2)   adoption of any part of one or more of these procedures
(3)   adoption of any other procedure which he considers to be appropriate
(4)   imposition of time limits

and as varied or amended by the arbitrator from time to time.

Rule 7 provides for a short hearing, rule 8 for 'documents only' and rule 9 for a full procedure. It can be seen from this that the arbitrator has considerable flexibility as to how to proceed and even subsequently to change the process. Here, as under the Act itself, there is the clear intention to provide whatever is considered the most appropriate procedure for settling the dispute and which provides more flexibility than generally contained within the 1988 JCT arbitration rules. Other rules deal with appointment of experts, costs and consolidation joinder of arbitrations.

ICE arbitration procedure

The Institution of Civil Engineers (ICE) recognised the need much earlier than JCT and produced the ICE Arbitration Procedure (England and Wales) (1983) which replaced the even earlier 1973 arbitration procedure. The civil engineering contracts produced by the ICE presuppose that arbitration under such contracts will be conducted in accordance with the relevant procedure. The ICE revised the procedure following the Arbitration Act 1996 and published the Institution of Civil Engineers Arbitration Procedure (1997), which for the purpose of their contracts is deemed to be an amendment to the 1983 procedure.

The Arbitration Procedure (1997) applies to the ICE Conditions of Contract 6th Edition, the ICE Conditions of Contract for Minor Works, the ICE Design and Construct Conditions of Contract and the NEC family of documents in England and Wales for arbitrations conducted under the Arbitration Act 1996. When these contracts are used in circumstances where the Arbitration Act 1996 is not applicable, these later rules may still apply because they amend the earlier rules. In these circumstances there is nothing to prevent the application of the 1983 rules but this should be made clear if that is what is required. The appendix to the various forms provides for the use of either the 1997 procedure or the CIMAR procedure.

The whole purpose of the arbitration procedures is to render arbitration a sensible option and to ensure it is not a pale imitation of High Court litigation and to take account of mandatory/non-mandatory provisions in the 1996 Act. Whether the new rules (1997) are more imaginative than the Arbitration Act 1996, with the latter being an improvement over the Arbitration Acts 1950–1979, remains to be seen. The way the 1996 Act is interpreted will be a major factor. At least the new legislation provides enhanced regulatory powers to an arbitrator under section 34, including limitation on discovery[21], dispensing with the strict rules of evidence[22], and adopting an inquisitorial procedure[23].

The pragmatism underlying the ICE Arbitration Procedure (1983) and the courts' wish to uphold it were demonstrated in *Christiani & Nielsen Limited* v. *Birmingham City Council* (1995). Under rule 1.2:

> 'the Notice to Refer [a dispute to arbitration] shall list the matters which the issuing party wishes to refer to arbitration. Where Clause 66 of the ICE Conditions of Contract applies the Notice to Refer shall also state the date when the matters listed therein were referred to the Engineer for his decision under Clause 66(1) and the date on which the Engineer gave his decision thereon or that he has failed to do so.'

The contractor's notice to refer failed to state the date on which the engineer had given his decision. Although clause 66(5)(a) of the ICE Conditions of Contract 5th Edition required a reference to arbitration to be 'conducted in accordance with the Arbitration Procedure (England and Wales) (1983)' this did not refer to the manner in which an arbitration was commenced. The judge was satisfied that the contractor's notice to refer, which complied with clause 66(3)(b) of the ICE conditions of contract, was a valid notice and was not invalidated by the procedural irregularity of failing to include all the matters required by rule 1.2 of the arbitration procedure.

In order to appreciate the developments in dispute resolution across the spectrum it is worth considering briefly the ICE Arbitration Procedure

(1997). This procedure is relatively short, consisting of 25 rules set out under parts A–J as follows:

- Part A. Objectives, Reference and Appointment
- Part B. Arrangements for the Arbitration
- Part C. Powers of the Arbitrator
- Part D. Procedures Before the Hearing
- Part E. Procedure at the Hearing
- Part F. Short Procedure
- Part G. Special Procedure for Experts
- Part H. Awards
- Part J. Miscellaneous.

Although these rules bear some similarity to the 1983 rules, there are some marked differences that take account of the Arbitration Act 1996, for example the detailed provisions on arbitrator's powers and the section on 'awards'. The influence of the Act is evident and now this has caught up somewhat, the procedure is more an elaboration than a provider of new extensive powers. Some of these powers would ordinarily be deployed by a judge and, in the context of arbitration, were contained within the earlier rules, many years before the general widening of an arbitrator's powers under the Arbitration Act 1996.

Under rule 6 the arbitrator determines the procedure to be adopted. The rules provide for three basic approaches (also incorporated in the 1983 rules): a full hearing, a special procedure for experts under rule 17 and the adoption of the short procedure under rule 15. The special procedure and the short procedure have a real part to play in the quest for cost-effective dispute resolution. The short procedure may be with a hearing or by documents only. However, the use of the special procedure for experts and the short procedure in either form requires the agreement or consent of the parties.

The short procedure is effectively a 'documents only' hearing where each of the parties puts together in the form of a file its statement of case. There is limited scope for oral submissions and questions from the arbitrator in rule 15.5, which states that the arbitrator shall fix a day to meet the parties, with his award to be made under rule 15.8 within 30 days following the conclusion of the meeting, with the parties or information received under rule 15.4 if there is no hearing. The arbitrator may also extend this period if it is reasonable to do so. Ordinarily under rule 16 the question of costs does not arise on a short procedure arbitration. The parties are liable equally for the arbitrator's fees and charges and are wholly responsible for their own costs. The parties are at liberty to inform the arbitrator that they no longer wish the short procedure to be adopted. The party wishing to depart from the short procedure will then become responsible for the whole of the arbitrator's fees and charges incurred prior to the giving of notice and will also be responsible for compensation to the other party for their costs incurred up to the date of notice. Clearly, this is an incentive to try and keep to the procedure adopted and to ensure early resolution of the dispute.

The special procedure is provided to meet the frequent occurrence in the construction industry of the dispute that involves matters of fact, which requires expert evidence. Many disputes in the construction industry (those that relate to complying with specifications, failure of materials, etc.) fall to be decided more on the facts than by an analysis of legal arguments. They call for sound technical assessment. For that reason the special procedure for experts should be of special interest on construction projects. At the heart of the special procedure is the reliance by each of the parties on a statement of case in file format. The principles are not dissimilar to those for the short procedure in that the arbitrator may view the works (rule 17.3) and have a hearing day when he meets the experts for them to make their submissions (rule 17.4). The one major distinction is that there is no overt mechanism for dispensing with the special procedure, once commenced, and the arbitrator is not required by rule 17.5 to provide his decision within a set period. Costs are dealt with in rule 18. Under the special procedure, the arbitrator has discretion with regard to the awarding of costs although, in the absence of specific agreement to the contrary, legal costs are excluded as a legitimate disbursement.

Also under this rule 6 the arbitrator can, together with the parties, determine whether or not to exclude the right of appeal to the courts. It will be interesting to see to what extent the exclusion will be used as it much depends on the advice given to the parties by their advisers.

The powers of the arbitrator are very wide and embrace the ability to rule on his own jurisdiction and to decide all procedural and evidential matters. These powers are contained within rule 7 and although the list of procedural and evidential matters is given, the power is not limited to those stated. The use of protective measures is also covered and gives both the arbitrator and the courts in cases of urgency the ability to make a range of orders that relate to the preservation of evidence, including taking photographs and the production of samples. Legal, technical or other assessors may be appointed by the arbitrator and may also seek such advice without the authority of the parties.

When not to arbitrate

Parties must identify those situations in which arbitration, and indeed ADR, are *not* the most suitable methods for resolving their dispute. In the past there has been a strong desire to find an alternative to arbitration but after recent developments arbitration may become more attractive. However, there will still be many circumstances when it may not be the best option or cannot be used.

Adjudication

Where an adjudication clause exists there may be either a right or obligation to use it, depending on the clause in the contract. If there is no such clause built into the contract and the HGCR Act applies, the parties are not obliged to go to adjudication; but if one party chooses to exercise this right, the other, albeit recalcitrant, party must submit to adjudication. Where a choice exists and one is concerned primarily with cash flow then this may be a good reason not to arbitrate.

Summary judgment

On occasions, a contractor will be advised that an application for summary judgment is preferable to arbitration. The obvious example is in cases where the contractor has not received certified or other monies when expected and his lawyer recommends summary judgment under RSC Order 14 (now CPR Part 24). However, case law has demonstrated that the mere issue of an architect's or engineer's certificate is no bar to cross claims being raised on behalf of the employer. This was not always the case. For a time an architect's certificate was considered as 'good as cash' and actions on unpaid certificates were treated in the same manner as those on dishonoured cheques. The authority for the dubious proposition that the employer had to pay up immediately and litigate cross-claims later was Lord Denning MR in *Dawnays Limited* v. *F.G. Minter Limited* (1971). The House of Lords overruled *Dawnays* in *Dawnays Limited* v. *F.G. Minter Limited* (1973) and the new orthodoxy accepted in *Mottram Consultants Limited* v. *Bernard Sunley & Sons* (1974). There are no *special* rules relating to cross claims in the construction industry. Given that building is not an exact science and creates claims legitimate and illegitimate, this severely limits the ability of contractors to obtain summary judgment payments. In the words of Lord Salmon in *Gilbert Ash*[24]:

> 'The [JCT] provisions relating to interim certificates as a rule ensure a steady cash flow in normal conditions... When, however a bona fide dispute arises, I do not think [they are] designed to put the plaintiff contractors or subcontractors in a fundamentally better position than any ordinary plaintiffs or the defendants in any worse position than any ordinary defendants.'

A plaintiff will obtain summary judgment where the defendant has no arguable defence to the claim. However, on a short hearing the courts cannot make a detailed investigation of the respective merits of claim and defence and the defendant will obtain leave to defend provided the defence appears arguable, however implausible it may be.

Traditionally, when a contractor made an application to the High Court for summary judgment the employer often retaliated with a cross application to transfer the proceedings to arbitration under section 4 (1) of the Arbitration Act 1950. Such a manoeuvre was to the advantage of the employer. Under the 1950 Act arbitration does not have as a general principle, any process akin to summary judgment in the High Court. An exception to this was the availability of an interim award under Rule 14 of the ICE Arbitration Procedure (England and Wales) (1983) and the principle set in one maverick decision, *Modern Trading Co. Limited* v. *Swale Building Limited* (1990). The Arbitration Act 1996 allows the arbitrator to provide interim relief, including provisional orders for the payment of money[25], but only if the parties have *agreed that the tribunal shall have power to order on a provisional basis any relief which it would have power to grant in a final award*. Otherwise the tribunal does not have such power.

The immediate problem faced by the courts on cross applications was to decide which summons should be heard first. In *Ellis Mechanical Services Limited* v. *Wates Construction Limited* (1976) the court was of the opinion that it was appropriate to hear the summary judgment application first. If it could be adequately shown that the defendant had no arguable defence to the plaintiff's claim, there was clearly no claim to refer to arbitration. *Hayter* v. *Nelson and Home Insurance Co.* (1990) subsequently created an element of confusion. Here the judge took the view that as soon as the summary judgment application was disputed there was a dispute or difference within the meaning of the arbitration clause, even if the particular dispute was capable of easy resolution. Therefore the judge felt obliged to transfer proceedings to arbitration. That said, *Hayter* is often viewed as a decision necessarily reached on the somewhat different wording of the Arbitration Act 1975 (which deals with international arbitrations) and not of specific relevance to domestic arbitration which was regulated principally by the Arbitration Act 1950. On occasions, the contractor succeeded on part of his claim as a summary judgment application with the remainder remitted to arbitration: *R.M. Douglas Construction Limited* v. *Bass Leisure Limited* (1990).

So how are applications to stay proceedings dealt with under the Arbitration Act 1996? Do summary judgement applications survive? Section 9 of the Arbitration Act 1996 states:

'9. – (1) A party to an arbitration agreement against who legal proceedings are brought (whether by way of claim or counterclaim) in respect of a matter which under the agreement is to be referred to arbitration may (upon notice to the other parties to the proceedings) apply to the court in which the proceedings have been brought to stay the proceedings so far as they concern that matter.

(2) An application may be made notwithstanding that the matter is to be referred to arbitration only after the exhaustion of other dispute resolution procedures.

(3) An application may not be made by a person before taking the appropriate procedural steps (if any) to acknowledge the legal proceedings

against him or after he has taken any steps in those proceedings to answer the substantive claim.

(4) On an application under this section the court shall grant a stay unless satisfied that the arbitration is null and void, inoperative or incapable of being performed.'

Superficially, section 9 (1) of the Arbitration Act 1996 has much in common with the earlier section 4 (1) of the Arbitration Act 1950. It is to be assumed applications under section 9 (1) will be made, as before, on summons and affidavit. Subsection (2) presumably refers to the possible intervention of adjudication, whether under the statutory scheme or one of the parties' own making. No change to the existing law appears to be made by subsection (3) but major changes appear to be intended by subsection (4). Under the 1950 Act, the court had discretion whether to stay proceedings to arbitration or not. The emphasis is now on the court being mandated to grant such a stay, with the wording of section 9 (4) closer to, albeit not identical, to that found in the Arbitration Act 1975. In *Halki Shipping Corporation* v. *Sopex Oils Limited* (1996) it was held, and latter upheld on appeal, that except for very limited circumstances, all disputes fell within the arbitration clause and had to be referred to arbitration. It is of course an issue whether, in a particular case, all disputes will fall within the arbitration agreement. Given the decision in Halki (the case law on section 9 mushrooming in the decisions in *Wenlands* v. *CLC Construction Limited* [1998] CLC 808, *Birse Construction Limited* v. *St David's Limited* (1999) CILL 1494 and *Jitendea Bhailbhai Patel* v. *Dilesh R Patel* (1999) CILL 1498) and that sections 5 and 6 of the Arbitration Act 1996 provide for a looser definition of a written agreement to arbitrate than that contained in section 32 of the Arbitration Act 1950, on the face of it, it will be much harder for the parties to evade the obligation to arbitrate, unless they otherwise agree.

The story would not stop here had it been for that fact that sections 85–88 of the Arbitration Act 1996 were not brought into force because they proved to be contrary to European law[26]. Clause 86, which was to have related to staying legal proceedings in relation to domestic arbitrations, provided greater opportunity to avoid a stay.

All this appears bad news for main contractors and subcontractors who might make summary judgement applications. Even section 39 of the 1996 Act, which leaves the parties free to agree that the tribunal shall have power to make provisional awards, is obviously unsatisfactory. The complication does not stop there. Section 108 of the Housing Grants, Construction and Regeneration Act 1996 may mean parties find themselves left just with adjudication as their instant remedy, even if the obligation to arbitrate is deleted from the parties' contract. The desire to give prompt effect to Latham principles via statutory adjudication also seems to have generated uncertainty in the law. It will take some time to decide where litigation, statutory adjudication, particularly in the form of the Secretary of State's Scheme for Construction Contracts, and arbitration fit together.

Will a party be denied a summary judgement application because his immediate remedy is statutory adjudication? The discussion of that begins another story.

Interim payments

Sometimes, in addition to an application for summary judgment under RSC Order 14 (CPR Part 24), or as an alternative to such an application, a plaintiff who feels he has a strong case for the early release of monies prior to any trial, has applied to the court for an interim payment. Such applications are made under RSC Order 29 (CPR Part 25). Applications for an interim payment will be successful where:

- the defendant has admitted liability but there is no agreement as to the precise level of quantum to be paid, *or*
- judgment has been entered for the plaintiff with damages to be assessed, *or*
- the court is satisfied that, if the action proceeded to trial, the plaintiff would be given judgment for a substantial amount.

Interim payments can be difficult to obtain in construction cases. As the prosecution of claims and their defence via counterclaims etc. is something of a mystery, plaintiffs often have an uphill battle in establishing any entitlement to an interim payment. In one case, *Crown House Engineering Limited* v. *Amec Projects Limited* (1989), the Court of Appeal held that an interim payment was possible 'to the extent that a claim, although not actually admitted, can scarcely be effectively denied'. *Shanning International Limited* v. *George Wimpey Limited* (1988) held that the court must be satisfied that if the action proceeded to trial the plaintiff would more likely than not obtain judgment for a substantial sum. The court must make a realistic assessment of any alleged rights of set-off or counterclaim which the defendant puts forward. Only then can the discretion be exercised in the plaintiff's favour. Again, interim payment applications suffer at the hands of section 9 of the Arbitration Act 1996.

Chapter 3
The Growth of ADR in the United Kingdom, Australia and Hong Kong

ADR in the UK

Unlike litigation, the impact of which is somewhat easier to analyse with the availability of national data on the number of referrals to the County Courts and the High Court, the extent to which other methods of dispute resolution have been used is difficult to assess. One problem ADR shares with arbitration is the lack of a single point of organisational control in the UK. In the same way that arbitrators may be appointed by the Chartered Institute of Arbitrators, the RICS, the RIBA, or a number of other appointing bodies, the growth of ADR depends on the efforts of a number of bodies which at times appear to be in competition with each other rather than merely complementary.

Perhaps the three most prominent promoters of ADR are the Academy of Experts (formerly the British Academy of Experts), CEDR (the Centre for Dispute Resolution) and the lawyer-led ADR Group, which is associated with one of the early providers of ADR services, IDR Europe Limited. What emerges from all such ADR providers is the apparent lack of hard data on the use and growth of ADR in the UK.

According to its own *Members Handbook*[1] 'the British Academy of Experts has been established to promote the better use of experts, to ensure that the standard of excellence already achieved is maintained and developed and to facilitate the efficient resolution of disputes'. The Academy, which was formed in 1987, now has a large membership, primarily drawn from the ranks of practising expert witnesses. The Academy is prominent in providing mediation training services to existing and potential third party neutrals. Section 3 of the *Members Handbook* includes a thumbnail sketch of mediation and how the Academy can assist in setting up mediation hearings.

Another prominent ADR provider and the one which appears to have enjoyed the greatest level of press coverage over the last few years or so is the Centre for Dispute Resolution which, with backing from the Confederation of British Industry, was formed in 1990. Its basis is somewhat different from the Academy; it has a number of member organisations drawn from commerce and industry as well as support from a large number of City and regional law firms. It has many similarities with the Center for

Public Resources in New York. However, like the Academy, CEDR is an important training organisation and appointer of mediators. In addition, to develop the use of mediation in particular market sectors, CEDR has set up specialist working groups, including the construction industry working group. This brings together construction lawyers, construction professionals and representatives drawn from contracting and client organisations and its purpose is to provide a forum in which developments are discussed, initiatives taken and information generally disseminated. CEDR has also been active in providing seminars and presentations in various parts of the UK, often in conjunction with the Confederation of British Industry, to increase the general level of awareness of ADR among businessmen.

The more cynical onlooker may allege that ADR has simply not taken off in the UK and it is difficult to point to firm data that clearly demonstrate an appreciable rise in activity since 1990. According to *The Lawyer*[2], ADR is the method of dispute resolution for 5% of construction disputes, negotiation for 13%, arbitration for 38%, adjudication for 1% and litigation for 43%. The reliability of such figures is obviously open to question. However, research data[3] indicate that ADR is only used in approximately 4% of construction disputes and lawyers are seen as a major impediment to its future development. The information directly available from the major ADR providers is again somewhat imprecise.

CEDR claims to conduct approximately 1–1.5 mediations per week, with the value of claims being anywhere between a few thousand pounds and fifty million pounds. After CEDR was established in 1990, approximately 1000 disputes were referred in about 5 years to the value of £1.5 billion, although the actual number of disputes resulting in full mediations is unavailable. CEDR has also had some experience of mini-trials and Med-Arb. For instance, CEDR was involved in one £200 million mini-trial as special adviser. According to the Academy of Experts, the level of activity remains relatively low, with not all referrals resulting in mediation hearings.

Perhaps the most positive message is that received from the ADR Group. A spokesman for the Group considered that there had been a considerable change in the public perception of ADR. In 1990 the Group received referrals for mediation at the rate of approximately three per month, of which one, on average, related to the construction industry. At that time, only one case would generally proceed as far as a full mediation. Many cases failed to reach a mediation hearing simply because one or other of the participants did not understand ADR, refused to participate or was committed to arbitration. In the period to 1993 there was a steady growth, with the number of referrals doubling to about six per month. The percentage of those relating to the construction industry remained approximately the same. The number of cases which reached a mediation hearing increased from about one-third to about one-half of the cases referred, although the number of construction disputes referred to a hearing did not increase. Apparently, the period from mid 1994 until mid 1995 saw a considerable change, with approximately five or six cases referred on average per week. Of those, at least 60% actually resulted in a mediation hearing and the success rate in those cases was at least 90%.

During the period to 1995, there was a corresponding increase in the value of claims referred to mediation. In 1990, for what was then an untried technique, disputes with a value of more than £100,000 were rarely referred. The ADR Group is now handling cases ranging from £50,000 to £2.4 million with the number of cases at around the £1 million mark becoming more common. It is in multi-party cases that there has been the greatest desire to take cases to mediation, given the high costs of litigation, the complexity of multi-party proceedings and the length of trials – several weeks or more. Many modern commercial parties choose not to live with the litigation lottery.

A further growth area for mediated settlements has been in professional negligence claims which may involve architects, surveyors or engineers. One case handled by the ADR Group related to four contractors and an architect in dispute over their liability. Importantly, not one of them alleged that the work carried out on behalf of the client was satisfactory. The mediation was dealt with in two parts. First, there was a hearing to establish agreement between the parties as to their degree of liability and second, a subsequent hearing to mediate between the plaintiff and the contractors and architects in regard to quantum.

Although the Chartered Institute of Arbitrators (CIArb) is primarily concerned with the promulgation of arbitration, nevertheless when ADR began to develop momentum after 1990, the Chartered Institute produced a number of initiatives of its own. That said, and although the Chartered Institute's own journal, *Arbitration*, has been instrumental in continuing the debate about the problems of arbitration and the option of ADR, the Chartered Institute has not been a prime promoter of ADR in the UK.

The growth of other methods of dispute resolution, related to ADR, has also been slow. The ICE Conciliation Procedures 1988 and 1994 bear this out. Up to 1994 the President of the ICE appointed approximately three conciliators a year, but since the ICE 6th Edition largely replaced the earlier 5th Edition there has been little growth in the use of conciliation. In 1995 conciliations were running at about 60 per year, of which ten/twelve per year relied on ICE appointed conciliators. Twenty/twenty five of the conciliations relate to the ICE Minor Works Form of Contract. The best available statistics suggest that party-appointed conciliators outnumber presidential appointments by 5/4:1.

One well known conciliator/arbitrator indicated that in the period to 1995 he had been engaged in six conciliations with values in the range of £40,000 to £2 million. At least one of those conciliations involved a local authority that indicated that it required a 'recommendation'. The rationale was that local authorities needed to satisfy their auditors, who would always react badly to mediation (the whiff of horse trading), whereas a conciliator's recommendation had the advantage of being a contractual response. The same conciliator also felt that, at least as far as civil engineering was concerned, the disputatious nature of the industry could be overstated. Apparently there is one arbitration for every £250 million worth of engineering work carried out in the UK.

The point relating to local authority auditors is well recognised in the USA and the Administrative Dispute Resolution Act was passed in 1990 to expand the use of ADR by federal agencies. This Act permitted such agencies to resolve all types of dispute, including construction disputes by ADR. However, things did not run that smoothly and further legislation and an amendment to this Act were made in 1996 to overcome shortcomings. The new legislation is expected to increase the use of ADR in the arena of public bodies, especially small claims under the Contract Disputes Act.

Many now consider ADR an acceptable dispute management tool and CEDR has helped to set up mediation schemes for organisations as disparate as the Department of Health, Building Employers Confederation, Computing Services and Software Association and the Institute of Grocery Distribution.

By contrast, support for ADR from professional organisations within the construction industry has been somewhat patchy. For instance, the Construction Industry Council (CIC) has indorsed ADR and in January 1993 the National Joint Consultative Committee for Building (NJCC) produced its Guidance Note 7, *Alternative Dispute Resolution*. According to the Guidance Note, [its purpose] 'is to describe the most popular methods of ADR, to suggest when they may be appropriate and when they should not be used and to look at the benefits and risks of the process'. The Guidance Note is extremely short, running to some four pages, but it is an extremely helpful brief overview of the main forms of ADR – mediation, executive tribunal (mini-trial) and non-binding adjudication.

The Guidance Note also examines when ADR can be used, when it is inappropriate, the problem of confidentiality and the advantages and disadvantages of the system. The section 'When ADR is inappropriate' suggests 'when auditors or others require an "imposed" decision'. It is often a source of frustration to those who promote ADR that Government departments, local authorities and other bodies which profess a need to maintain public accountability always raise the spectre of the auditor as a ground for not adopting ADR techniques. Such people either forget or disregard the fact that most legal actions are not resolved in a forensic manner but are settled by way of a belated 'horse deal' at some stage prior to or, at worst, on the steps of the court building. Rather than sound such a negative note in the Guidance Note with regard to the perceived problem of public auditors, it would have been more helpful if the NJCC had emphasised the need to carry out an educative process among local authorities and other similar organisations. The reluctance of local authorities to change their approach, other than with extreme caution, is demonstrated by the desire of many for a recommendation from the conciliator whenever they agree a reference to conciliation under the ICE Conciliation Procedure.

Among professionals, some of the greatest opposition to the growth of ADR in the UK has traditionally come from the RIBA, although an RIBA architect/client conciliation scheme was developed. The RIBA remains extremely conscious of the relative fragility of its members' position as contract administrators with the rise of construction procurement methods that challenge the traditional authority of architects. Many architects think back with nostalgia to a time when the architect's discretion was unchallenged and the architect truly saw himself as the independent contract administrator. With the making of other inroads into the architect's traditional role, members of the RIBA have been reluctant to see the architect's position further eroded by apparent contractual challenges to his supremacy as contract administrator, although the architect will now need to adjust to statutory adjudication. The RIBA did, however, introduce a limited scheme for the adjudication or conciliation of disputes between clients and its members.

The JCT, of which the RIBA is a pivotal member, has after much initial reluctance now seemingly shown support for ADR by publishing its Practice Note 28, *Mediation in a Building Contract or Sub-Contract Dispute*[4]. The Practice Note makes the obvious point that disputes can be resolved by either litigation, arbitration, adjudication or agreement. Mediation is a method of assisted negotiation. Paragraph 2 sounds the 'health warning' that mediation may not necessarily be suitable where the definitive intervention of the courts is necessary. Such circumstances might include those where an injunction is required, a public hearing is appropriate, or one or other of the parties has no genuine interest in resolving the dispute. Although paragraph 2 provides useful bullet points, it does not fully consider those situations where mediation is unsuitable. This task is better achieved in the NJCC Guidance Note.

Quite usefully, paragraph 4 of Practice Note No. 28 indicates that mediation is only one alternative and a short order arbitration under rule 7 of the JCT Arbitration Rules 1988 may be more appropriate. The Practice Note indicates that mediation does not prevent a party commencing or, if already commenced, continuing with arbitration or litigation. Unlike some commentators on ADR, who are of the opinion that mediation, unlike adjudication, should not be available during the currency of a contract, the JCT suggests that mediation may occur pre-practical completion of the works. Paragraph 6 states that the parties should proceed with the contract as if there were no disputes during the period when the mediation occurs. Paragraph 7, apart from stating that the procedure set out in the Practice Note will have to be amended whenever there are multi-party proceedings, provides no further guidance on how to deal with those disputes.

There is something of an understatement in paragraph 8, which states that 'The Architect, Quantity Surveyor or other appointed consultant will usually need to be involved in the Mediation so that their comments are available to the parties'. The professional team is obviously essential. Many, if not all, disputes arise from the way in which the professional team has exercised its various discretions during the pre-contract and construction phases of a particular project. Again, paragraph 9 deals somewhat mutedly with the issue of the decision or discretion of the architect or quantity surveyor being varied or negated by any mediation settlement. The paragraph simply states that a copy of the agreement should be given to the appropriate consultant. Then, by way of complete understatement, the paragraph concludes: 'such terms might affect or be considered by them to affect their professional responsibilities under their agreement with the Employer'.

Rather usefully, the Practice Note concludes with some draft agreements. These are a mediation agreement, an agreement appointing a mediator and an agreement following resolution of a dispute after mediation. The purpose of the final document is to address one of the biggest criticisms of ADR, the fact that it is non-binding. The best response is to set out the settlement in a document which can be used to support enforcement of the decision. By agreement of the parties, an ADR settlement may, following non compliance with its terms, be deemed capable of enforcement as a judgment of the High Court or, in the alternative, as an arbitration award registrable for execution under an Arbitration Act. What the parties should wish to avoid in their final agreement is any possibility that the courts or an arbitrator may be able to open up and re-try the matters in dispute because of a loosely worded settlement agreement.

ADR in Australia

In Australia, ADR has comprised the following methods of dispute resolution:

- mediation
- conciliation (the term here is generally used to refer to compulsory mediation in the context of court or tribunal proceedings)
- adjudication
- expert appraisal
- mini-trial or case presentation.

The growth of all forms of ADR has been rapid over the past decade, as illustrated by the courts' consideration of points arising from ADR processes. In part, this is due to a reaction by the community against the delays and complications of going to law, but more particularly the Federal and State Governments have become concerned at the spiralling cost of providing legal services and assuring the provision of justice. Compulsory and voluntary ADR (particularly mediation and case appraisal) have become part of the procedure in many Australian courts. The Australian construction industry is one of the most identifiable sectors of the Australian economy where a noticeable trend away from litigation has taken place. In May 1990 a joint working party, containing representatives from all the major groups within the industry, published the results of its investigation into strategies for the reduction of claims and disputes in the construction industry. Its report, *No Dispute,* made wide ranging recommendations for changes in the industry to improve practices and reduce claims and disputes. *No Dispute* summarised the prevailing problems with handling claims in the Australian construction industry in the following way:

- Although most forms of contract contained provisions for dispute resolution, too often resolution *at the working level* had been frustrated, resulting in the use of more formal adversarial processes, such as arbitration and litigation.
- Generally, a satisfactory outcome in litigation and arbitration was precluded by the excessive costs associated with them and the delays arising from the multiplicity of disputes awaiting hearing in the courts and before arbitrators.
- A consequential cost arose from the attrition and mobility of staff within the industry who possessed the hands on knowledge of particular disputes.

The report recommended the formal introduction of ADR procedures into construction contracts and suggested a pro forma dispute clause. This included compulsory negotiation between the parties and an obligation to endeavour to agree an appropriate ADR mechanism to resolve the dispute before it could be taken to arbitration. The full text of the clause is:

'(i)   It is the intention of the parties that in the event of a dispute, difference, controversy or claim arising out of or relating to the performance of the contract or the breach, rectification, termination, frustration or invalidity thereof, (hereinafter referred to as "the dispute"), every endeavour shall be made to resolve the matter on its merits by negotiation. The parties shall attend at least one meeting to discuss the matter at issue, as a condition precedent to commencing any other proceeding in respect of the dispute. If the dispute cannot be resolved by negotiation as aforesaid the parties shall confer in order to ascertain whether they agree that the dispute shall first be subject to the process of conciliation, mediation, appraisal or mini-trial, or such other alternative dispute resolution process as may be appropriate in the circumstances of the dispute and if they so agree, the dispute shall be referred to such process.

(ii)   In the event that the dispute cannot be resolved in accordance with the procedure set out in paragraph (i), or if at any time either party reasonably considers that the other party is not making reasonable efforts to resolve the dispute, a notice may be issued to the other party requiring that the dispute be referred to arbitration.

(iii)   Arbitration shall be effected by a single Arbitrator in accordance with and subject to the Institute of Arbitrators Australia Rules for the Conduct of Commercial Arbitration. The Conciliator, Mediator,

Appraiser or neutral Advisor appointed by the parties, shall not be appointed as Arbitrator nor may that person be called by either party in the arbitration unless both parties agree in writing.

(iv)   The parties and the Arbitrator shall meet to discuss and develop procedures appropriate to expedite the conduct of the arbitration and the parties shall co-operate with the Arbitrator in expeditious conduct of the arbitration.

(v)   The cost of the submission, reference and award, together with the apportionment thereof shall be at the discretion of the Arbitrator.'

The 'No Dispute' recommendations have been, in principle at least, embraced enthusiastically by the Australian construction industry. For example, the subsequently drafted Australian Standard AS2124–1992 and AS4300–1995 (see Appendix 1 for full details of the clauses) both specifically require negotiations between the parties and an attempt to reach agreement by a suitable ADR process.

Bespoke construction contracts now commonly contain similar provisions to those contained in these standard contracts, requiring at least one form of ADR to be attempted. Perhaps more significantly, even if contracts do not contain ADR provisions, there is a clear tendency for disputes to be referred to mediation and case appraisal before either arbitration or litigation are instigated. The rise of ADR has been accompanied by a prolific growth in the range of organisations and firms established to facilitate ADR. Examples include the National Disputes Centre (NDC), Lawyers Engaged in Dispute Resolution (LEADR), and the Australian Commercial Disputes Centre (ACDC). Industry based organisations, most notably the Master Builders Association, have also established mediation and case appraisal schemes. The Australian Institute of Arbitrators amended its constitution in 1995 to enable it to promote, teach and handle all forms of ADR and has published rules for the conduct of commercial conciliations

The continued growth of ADR in Australia, as elsewhere, depends in part on judicial support. There are two aspects in particular where the courts' views are critical:

- Will the courts force reluctant parties to mediate (or participate in other forms of ADR) where they have entered into a contract to do so?
- Will the courts respect the privacy and confidentiality of ADR processes, particularly where they fail to resolve disputes?

Both these issues have been the subject of judicial consideration in Australia and are discussed below because they impact directly on the development of ADR. Further discussion of the courts' consideration of ADR clauses is provided in Chapter 8.

In *Allco Steel (Queensland) Pty Ltd* v. *Torres Strait Gold Pty Ltd* (1990), the court examined a contract which provided for conciliation between the parties prior to litigation. When a dispute arose, one of the parties consistently demonstrated what the judge called a 'complete lack of conciliatory spirit'. Although the court found that one of the parties had committed 'a clear breach of [its] obligations to conciliate'[5], it refused to grant a stay of the legal proceedings until the conciliation process had been completed. In doing so, the court relied on the doctrine that the jurisdiction of the courts cannot be ousted. The court also found it material that one party had adopted an extremely negative view of conciliation:

> 'In other words, notwithstanding what I perceive to be a clear breach of the obligations to conciliate on the part of the plaintiff, the doctrine that the jurisdiction of the court cannot be ousted dominates any other principle that would require the plaintiff to honour its contractual obligations that might arise under Clause 4.5.6. An appeal was made to the inherent jurisdiction of the court to grant a stay, the condition precedent to the accruing of a cause of action not having been met, namely bona fide conciliation. In my view, even if such relief was open, this discretionary relief must be refused as it is abundantly clear that the parties have taken up positions which effectively rule out the possibility of compromise and conciliation, the plaintiff by its assertion that 'discovery' does not lie pursuant to Clause 4.5.3 and the defendants by their insistence on such process as a precondition of negotiations.'

Subsequent decisions in New South Wales, notably *Hooper Bailie Associated Limited* v. *Natcon Group Pty Ltd* (1992) and *Elizabeth Bay Developments Pty Ltd* v. *Boral Building Services Pty Ltd* (1995) did not follow *Allco*. In the latter case the court refused to enforce a mediation agreement because it lacked certainty, but endorsed the view that an agreement to conciliate or mediate is enforceable[6]. These decisions demonstrate:

- It is an abuse of process to commence proceedings without complying with a contractual provision for conciliation.
- The exercise may not be futile – an independent third party (i.e. the mediator) may be able to overcome the parties' resistance to ADR.
- An agreement to mediate or conciliate is not merely an 'agreement to negotiate' – it is an agreement to participate in a process.[7]

Confidentiality may be an inducement to use ADR rather than litigation so whether proceedings remain confidential could be important. The confidentiality of the mediation process was considered in the long running proceedings in *AWA Limited* v. *Daniels & Others* (1992).

At an interlocutory stage, Rolfe J of the Supreme Court of New South Wales was called on to decide whether or not information disclosed by one of the parties during an aborted mediation could later be used by one of the other parties. One of the defendants had issued a notice to produce, requiring production of documents, the existence of which had been disclosed in the course of the mediation between the parties.

The court refused to set aside the notice to produce, allowing the defendant to utilise facts obtained during the mediation. Rolfe J drew a close analogy between mediation and settlement negotiations and relied on the decision of the High Court in *Field* v. *Commissioner of Railways (NSW)* (1957) which narrowed the privilege pertaining to settlement negotiations such that it did not preclude a party leading evidence on facts ascertained during the negotiations.

Later, in the same case and in the context of an objection to the tender of the same documents, Rogers CJ overruled the objection but said:

'It is of the essence of successful mediation that parties should be able to reveal all relevant matters without an apprehension that the disclosure may subsequently be used against them ... were the position otherwise, unscrupulous parties could use and abuse the mediation process by treating it as a gigantic, penalty free discovery process...'

One relevant factor in both judges' decisions appears to be that the defendant's solicitor was aware of the possible existence of the material sought before the mediation and that the documents were probably discoverable if their relevance could be established. Nonetheless, it appears unsafe to assume, at least in Australia, that parties cannot utilise information obtained through mediation in subsequent court proceedings. The issue does not appear to have subsequently come before the courts although many commentators predict legislative intervention[8].

A fairly recent development in the area of dispute resolution is the amendment of section 27 of the Commercial Arbitration Act[9] to permit the parties to an arbitration agreement to seek settlement of disputes 'by mediation, conciliation or similar means' and may authorise an arbitrator to act as mediator, conciliator or other 'non-arbitral intermediary'. The section also requires the arbitrator, in performing that role, to be bound by the rules of natural justice and enables an arbitrator who has fulfilled that role to resume as arbitrator if the dispute is not settled.

Although the amendments to section 27 now make it clear that an arbitrator, having initially tried to facilitate a settlement, can continue as arbitrator without being attacked for bias, in practice many arbitrators (and also many parties to arbitrations) are extremely reluctant to participate in a process where one (albeit neutral) party acts both as a mediator and an arbitrator (see discussion on Med-Arb in Chapter 7).

The full text of section 27 is:

Commercial Arbitration Act 1984 No. 160
'Part 3 – Conduct of Arbitration Proceedings 27.
Settlement of Disputes Otherwise than by Arbitration
27.   Settlement of disputes otherwise than by arbitration
    (1)   Parties to an arbitration agreement:
        (a)   may seek settlement of a dispute between them by mediation, conciliation or similar means; or

(b)   may authorise an arbitrator or umpire to act as a mediator, conciliator or other non-arbitral intermediary between them (whether or not involving a conference to be conducted by the arbitrator or umpire),

whether before or after proceeding to arbitration, and whether or not continuing with the arbitration.

(2)   Where:

(a)   an arbitrator or umpire acts as a mediator, conciliator or intermediary (with or without a conference) under sub-section (1); and

(b)   that action fails to produce a settlement of the dispute acceptable to the parties to the dispute,

no objection shall be taken to the conduct by the arbitrator or umpire of the subsequent arbitration proceedings solely on the ground that the arbitrator or umpire had previously taken that action in relation to the dispute.

(3)   Unless the parties otherwise agree in writing, an arbitrator or umpire is bound by the rules of natural justice when seeking a settlement under sub-section (1).

(4)   Nothing in sub-section (3) affects the application of the rules of natural justice to an arbitrator or umpire in other circumstances.

(5)   The time appointed by or under this Act or fixed by an arbitration agreement or by an order under section 48 for doing any act or taking any proceeding in or in relation to an arbitration is not affected by any action taken by an arbitrator or umpire under sub-section (1).

(6)   Nothing in sub-section (5) shall be construed as preventing the making of an application to the Court for the making of an order under section 48.'

ADR in Hong Kong

Despite recent economic turmoil in the region and political change, Hong Kong continues to be a great centre of construction activity. This, together with cultural differences in the Asia/Pacific Rim, has meant a greater enthusiasm for looking at innovative methods of dispute resolution. In addition, in the Asia/Pacific Rim there may be a cultural preference in favour of mediation as opposed to adjudicative methods of dispute resolutions. According to an American lawyer, Donahey:

'In various Asian countries, there is a profound societal philosophical preference for agreed-upon solutions. Rather than a cultural bias towards "equality" in relationships, there exists an intellectual and social predisposition towards a natural hierarchy which governs conduct in interpersonal relations. Asian cultures frequently seek a "harmonious" solution, one which tends to preserve the relationship, rather than one which, while arguably, factually and legally "correct", may severely damage the relationship of the parties involved.'[10]

Donahey identified the Chinese approach as being in keeping with traditional Confucianism[11]:

'Within traditional Confucianism, going back thousands of years, there is a concept known as *li* which concerns the social norms of behavior within the five natural status relationships: emperor and subject, father and son, husband and wife, brother and brother, or friend and friend. *Li* is intended to be persuasive, not compulsive and legalistic, a concept which governs good conduct and is above legal concepts in societal importance. The governing legal concept, *fa* is compulsive and punitive. While having the advantage of legal enforceability, *fa* is traditionally below *li* in importance. The Chinese have always considered the resort to litigation as the last step, signifying that the relationship between the disputing parties can no longer be harmonized. Resort to litigation results in loss of face, and discussion and compromise are always to be preferred. Over time the concept of *fa* and *li* have become confused and the concept of maintaining the relationship and, therefore, face, has become part of the Chinese legal system.'

Donahey concluded his article by referring to a similar position that exists under other Asian legal systems, including the Korean one.

Traditionally, Hong Kong did not deviate from the English model. Statutory control of Hong Kong arbitration, found in the 1963 Arbitration Ordinance, was identical to the Arbitration Act 1950. Hong Kong enacted a new arbitration ordinance in 1982[12], which, although modelled on the Arbitration Act 1975, was significantly different. One of the principal differences from English practice was specific recognition of conciliation. In 1985, the provision of non court based dispute resolution services began to be co-ordinated by the establishment of the Hong Kong International Arbitration Centre (HKIAC), set up in part to consolidate Hong Kong's pivotal role in Asian trade. HKIAC publishes a helpful booklet[13] which emphasises HKIAC's dedication to the promotion of mediation as a dispute resolution technique and includes a standard mediation clause for domestic (i.e. Hong Kong) mediation[14]. The recommended clause reads as follows:

'Any dispute or difference arising out of or in connection with this contract shall first be referred to mediation at Hong Kong International Arbitration Centre (HKIAC) and in accordance with its Mediation Rules. If the mediation is abandoned by the mediator or is otherwise concluded without the dispute or difference being resolved, then such dispute or difference shall be referred to and determined by arbitration at HKIAC and in accordance with its Domestic Arbitration Rules.'

What was significant about the 1982 Arbitration Ordinance was its radic-
alism. The Law Reform Commission of Hong Kong had recommended
inclusion of provisions for conciliation in the Arbitration Ordinance in its
1981 Report[15]. These provisions were set out in the following terms:

'2.   The principal Ordinance is amended by adding, after Part 1, the
following Part –

Part A
CONCILIATION

Appointment
of Conciliator.

2A   (1)   In any case where an arbitration agreement provides for the
appointment of a conciliator by a person who is not one of the
parties and that person refuses to make the appointment or does
not make it within the time specified in the agreement or, if no
time is so specified, within a reasonable time not exceeding 2
months of being informed of the existence of the dispute, any
party to the agreement may serve the person in question with a
written notice to appoint a conciliator (and shall forthwith serve
a copy of the notice on the other parties to the agreement) and if
the appointment is not made within 7 clear days after service of
the notice the Court or a judge thereof may, on the application of
any party to the agreement, appoint a conciliator who shall have
the like powers to act in the conciliation proceedings as if he had
been appointed in accordance with the terms of the agreement.

(2)   Where an arbitration agreement provides for the appointment
of a conciliator and further provides that the person so
appointed shall act as an arbitrator in the event of the conciliation proceedings failing to produce a settlement acceptable
to the parties –

(a)   no objection shall be taken to the appointment of such
person as an arbitrator, or to his conduct of the arbitration
proceedings, solely on the ground that he had acted previously as a conciliator in connexion with some or all of the
matters referred to arbitration;

(b)   if such person declines to act as an arbitrator any other
person appointed as an arbitrator shall not be required first
to act as a conciliator unless a contrary intention appears in
the arbitration agreement.

(3)   Unless a contrary intention appears therein, an arbitration
agreement which provides for the appointment of a conciliator
shall be deemed to contain a provision that in the event of the
conciliation proceedings failing to produce a settlement acceptable to the parties within 3 months, or such longer period as the
parties may agree to, of the date of the appointment of the
conciliator or, where he is appointed by name in the arbitration

> agreement, of the receipt by him of written notification of the existence of a dispute the proceedings shall thereupon terminate.
>
> (4)   If the parties to an arbitration agreement which provides for the appointment of a conciliator reach agreement in settlement of their differences and sign an agreement containing the terms of settlement (hereinafter referred to as the "settlement agreement") the settlement agreement shall, for the purposes of its enforcement, be treated as an award on an arbitration agreement and may, by leave of the Court or a judge thereof, be enforced in the same manner as a judgement or order to the same effect, and where leave is so given, may be entered in terms of the agreement.'

The provisions set out above contained a number of significant features which are not even contemplated for inclusion in English law and practice. First, a recalcitrant party could be obliged to conciliate; the court had a statutory right and power to uphold conciliation agreements. Second, the ordinance endorsed the principle of Med-Arb. Third, a finite period was set for completion of the conciliation phase. Fourth, statute maintained that any settlement agreement should, for the purposes of enforcement, have the same status as an arbitration award and be enforceable through the courts.

The matter of conciliation was considered further by the Law Reform Commission of Hong Kong[16] and its recommendations included in the Arbitration Ordinance in 1989[17]. The relevant changes are set out below. Section 2A (1) was amended, Section 2A (4) repealed and new Sections 2B and 2C inserted.

> 'Appointment
> of conciliator
> 2A   (1)   In any case where an arbitration agreement provides for the appointment of a conciliator by a person who is not one of the parties and that person refuses to make the appointment or does not make it within the time specified in the agreement or, if no time is so specified, within a reasonable time of being requested by any party to the agreement to make the appointment, the Court or a judge thereof may, on the application of any party to the agreement, appoint a conciliator who shall have the like powers to act in the conciliation proceedings as if he had been appointed in accordance with the terms of the agreement. (*Amended 64 of 1989 s.4*)

> Power of arbitrator
> to act as conciliator
> 2B   (1)   If all parties to a reference consent in writing, and for so long as no party withdraws in writing, his consent, an arbitrator or

umpire may act as a conciliator.

(2) An arbitrator or umpire acting as conciliator:
 (a) may communicate with the parties to the reference collectively or separately;
 (b) shall treat information obtained by him from a party to the reference as confidential, unless that party otherwise agrees or unless subsection (3) applies.

(3) Where confidential information is obtained by an arbitrator or umpire from a party to the reference during conciliation proceedings and those proceedings terminate without the parties reaching agreement in settlement of their dispute, the arbitrator or umpire shall, before resuming the arbitration proceedings, disclose to all other parties to the reference as much of that information as he considers is material to the arbitration proceedings.

(4) No objection shall be taken to the conduct of arbitration proceedings by an arbitrator or umpire solely on the ground that he had acted previously as arbitrator in accordance with this section. (*Added 64 of 1989 s.5*)

Settlement agreements

2C If the parties to an arbitration agreement reach agreement in settlement of their dispute and enter into an agreement in writing containing the terms of settlement (the "settlement agreement") the settlement agreement shall, for the purposes of its enforcement, be treated as an award on an arbitration agreement and may, by leave of the Court or a judge thereof, be enforced in the same manner as a judgement or order to the same effect and, where leave is so given, judgement may be entered in terms of the agreement. (*Added 64 of 1989 s.5*)'

In the interpretation of the arbitration ordinance, the meaning of conciliation was extended to include mediation (*Added 75 of 1996 s.3*). A new sub-clause, which set out the objectives and principles of the Ordinance, was also introduced, together with a number of other additional sections relevant to the operation of arbitration. The new sub-clause said:

'2AA (1) The object of this Ordinance is to facilitate the fair and speedy resolution of disputes by arbitration without unnecessary expense.

(2) This Ordinance is based on the principles that –
 (a) subject to the observance of such safeguards as are necessary in the public interest, the parties to a dispute should be free to agree how the dispute should be resolved: and
 (b) the Court should interfere in the arbitration of a dispute only as expressly provided by this Ordinance. (*Added 75 of 1996 s.4*)'

The judiciary has played a prominent role in the promotion of ADR in Hong Kong. For instance, in April 1991 the chief justice appointed a committee to consider the desirability of a court annexed mediation scheme and related matters. The committee, which was chaired by Kaplan J, submitted its report to the Chief Justice in August 1993[18]. Although the report was subject to consideration by the Civil Court Users Committee it strongly recommended the use of court annexed mediation, although not subject to coercive powers to force parties to mediate if they did not so wish[19].

The Hong Kong courts' commitment to ADR is further evidenced by the *Construction List Practice Direction*[20]. This deals with the procedures to apply on interlocutory summonses which last over an hour. It includes the following:

'At the summons for directions the court shall be informed whether any and if so what attempt has been made to resolve the dispute or any part of it by mediation. This requirement does not entail disclosing the details of any mediation only the fact of it having taken place.'

The Hong Kong Government, a major procurer of construction work, has been active in the promotion of ADR. Relevant provisions are found in the standard form engineering and building contracts used by the Government[21]. Clause 86 states:

'86. (1) If any dispute or difference of any kind whatsoever shall arise between the Employer and the Contractor in connection with or arising out of the Contract or the carrying out of the Works including any dispute as to any decision, instruction, order, direction, certificate or valuation by the Engineer whether during the progress of the Works or after their completion and whether before or after the termination, abandonment or breach of the Contract, it shall be referred to and settled by the Engineer who shall state his decision in writing and give notice of the same to the Employer and the Contractor ... Such decision shall be final and binding upon the Contractor and the Employer unless either of them shall require that the matter be referred to mediation or arbitration as hereinafter provided. If the Engineer shall fail to give such decision for a period of 28 days after being requested to do so or if either the Employer or the Contractor be dissatisfied with any such decision of the Engineer then either the Employer or the Contractor may within 28 days after receiving notice of such decision, or within 28 days after the expiration of the said decision period of 28 days, as the case may be, request that the matter shall be referred to mediation in accordance with and subject to the Hong Kong Government Mediation Rules or any modification thereof for the time being in force.

(2) If the matter cannot be resolved by mediation, or if either the
Employer or the Contractor do not wish the matter to be
referred to mediation then either the Employer or the Contractor may within the time specified herein require that the
matter shall be referred to arbitration . . .'

The Hong Kong Government has also been involved in other ADR initiatives. First, in conjunction with the Hong Kong Institution of Engineers, the
Government produced in 1984 Mediation Service Rules which were superseded in 1989[22] and the Hong Kong Government Mediation Rules (1992
edition), administered by HKIAC[23]. In 1994, a mediation group of the
HKIAC was founded to promote mediation as a means of dispute resolution.
Neither of the sets of rules is a lengthy document. The first runs to some 21
rules and the second to 23 rules. The Mediation Service Rules consider the
definition of *Mediation* (rule 1), *Initiation of Mediation* (rule 3), *Appointment of
Mediator* (rule 5), *Conduct of the Mediation* (rules 7–17 inclusive), *Costs* (rules
18 and 19), *The Exclusion of any Med-Arb Possibility* (rule 20). Some of the more
important provisions are found in rules 7–17. Under rule 7 the mediator will
use his best endeavours to conclude the mediation within 42 days of
appointment. Rule 12 permits him to express preliminary views during the
course of the mediation and to seek third party guidance, as he sees fit. The
need for a supplementary agreement to reflect the terms of settlement (if
reached) is found in rule 13. The privacy and confidentiality of mediation is
reflected in rule 16. The mediation process, if unsuccessful, does not prejudice the parties' rights of action in any subsequent litigation or arbitration
(rule 17).

The more significant provisions in the Hong Kong Government Mediation
Rules are as follows. The parties may be contractually obliged to mediate
(rule 2) with mediation being possible if:

- either party to the contract disagrees with a decision of the engineer or
  architect, *or*
- the engineer or architect fails to make a decision within the requisite
  period.

Mediation is not mandatory under the rules; a party can within 28 days after
receipt of a request to mediate, indicate whether or not he is willing to
participate in the mediation. Conduct of the mediation is dealt with under
rules 7–18 inclusive. The provisions are broadly similar to those set out in the
Mediation Service Rules. In the absence of a satisfactory resolution of the
dispute (rule 14), the mediator first indicates his view orally to the parties.
This is without prejudice to his potential duty under rule 15 to provide a
report to the parties. A major distinction between the Mediation Service
Rules and the Government Mediation Rules is that rule 15 of the latter states
that the report will be prepared on a request from either party and is not
mandatory as under the former. In his report, the mediator sets out the facts

as he finds them, his opinion on the matters in dispute and proposals for settlement which he considers appropriate. Within a 28 day cooling off period, the parties may accept the mediator's proposed terms of settlement either in whole or in part. Under both mediation schemes (rules 19 and 20 respectively) if, in the mediator's opinion, the mediation has been initiated or conducted frivolously or vexatiously the mediator is empowered to make punitive awards of costs. Again, under the Government Mediation Rules as under the Mediation Service Rules, the possibility of Med-Arb is precluded (rule 21).

The current Government Mediation Rules have been the subject of review for a number of years. New rules were originally anticipated to be in effect before the end of 1996 but as yet they have not been published. The main purpose of the review is to remove those provisions which allow or require the mediator to express his views on the issues, or if the parties are unable to reach agreement, to submit a report recommending terms for settlement. Apparently, practice has shown these provisions not to have been conducive to encouraging negotiated settlements during the mediation.

One of the most substantial projects recently undertaken in Hong Kong relates to the new Hong Kong International Airport. The airport construction programme consisted of ten interlinked projects referred to as the Airport Core Programme (ACP). The ACP projects were the responsibility of the Hong Kong Government, which oversaw the infrastructure related contracts (excluding the airport railway) including the formation of new land, the construction of highways and a new town to serve the airport. The Airport Authority is responsible for all contracts relating to the airport with the Mass Transit Railway Corporation responsible for the airport railway contracts. All three procurers rely on a combination of dispute resolution mechanisms.

Under the Government ACP Form of Contract, disputes may go through four distinct stages: a decision of the engineer, mediation, adjudication and arbitration. On a dispute arising, either party may serve on the other and the engineer a notice of dispute, stating the nature of the dispute, a mechanism now well known to English contractors and contract administrators. Within 28 days of service, the engineer must either provide a decision or indicate that the original decision which led to the dispute was not his. If the matter is not resolved, the aggrieved party must within 28 days serve a request for mediation in accordance with the Government ACP Mediation Rules, 1992 Edition. Any mediation is conducted by a single mediator (rule 4) and will be founded on the request for mediation (rule 3.2.1) which contains 'a brief self-explanatory statement of the nature of the Dispute, the quantum in dispute (if any), and the relief or remedy sought'.

Although the appointment of a particular mediator can be rejected (rule 5), mediation cannot be excluded as a first stage dispute resolution mechanism. The conduct of the mediation is in accordance with rules 8–19 inclusive. A mediator may recommend appropriate settlement (rule 16). The question of costs and their resolution is set out at some length in rule 20. Significantly, rule 22 precludes the possibility of the mediator subsequently acting as an arbitrator on the dispute. As mediation is non-binding, the parties may need to consider moving to adjudication. The time at which the mediation option ceases to apply is quite difficult to ascertain. Under rule 8, the mediator is required to act 'as expeditiously as possible, but in any event the mediation process shall not continue for more than 42 days from the Mediation Commencement Date unless the parties agree otherwise'. Identification of the mediation commencement date under rule 6.1 is quite difficult. However, rule 17 makes it clear that the 42 day period referred to in rule 8 is qualified. It states 'the mediator may abandon the mediation whenever his judgment and further efforts at mediation would not lead to a settlement of the Dispute...'.

However, assuming that the parties are able to identify, or agree, the date on which the adjudication phase begins, any adjudication is carried out in accordance with the ACP Adjudication Rules (1992 Edition). Rule 1.2 makes it clear that the adjudicator cannot be empowered unless and until there has been a reference to mediation. Rules 2 and 3 deal with the commencement of the adjudication process and the appointment of an adjudicator. Rule 6 is of interest; rule 6.1 specifically states that an adjudicator 'shall have the widest discretion permitted by law to determine the procedure of the adjudication and to ensure the just, expeditious and economical determination of the Dispute...'. It is within his discretion to decide whether or not to have a hearing for the taking of oral evidence, whether to proceed on a 'documents only' basis and (rule 7) to regulate the calling of witnesses before him. The adjudicator is empowered under rule 7.3 to arrange for his own expert evidence. Wide ranging powers are provided to the adjudicator in rule 8 including, under rule 8.1.1, the power to examine any witness or carry out an inspection in the absence of the parties or their representatives.

To bring finality to the process, the adjudicator is required to make his decision within 42 days of the adjudication commencement date, subject to the parties' right to agree, in writing, a different period. The initial period of 42 days cannot be extended by more than 28 days. The adjudicator's decision is in writing and includes on monetary sums owed, a power given by rule 9.3 to award that interest be paid. The question of costs is dealt with in rule 10. Rule 11 does not empower the adjudicator to act subsequently as an arbitrator of the same or a related dispute. Adjudication is only available on questions of entitlement of 'one party to payment by the other pursuant to any provision of the Contract; and/or the Contractor to an extension of time...'.

Similar considerations arise in regard to the various contracts entered into by the Mass Transit Railway Corporation (MTRC). These contracts also made provision for mediation but not adjudication. As has been pointed out[24], the difficulty of imposing mandatory mediation arises from the MTRC form of contract. This includes the following clause[25]:

'No steps shall be taken in any reference to arbitration (with the exception of the formal appointment of the arbitrator) unless and until either party has first referred the dispute to mediation by serving on the other party a request for mediation in accordance with Clause 103.7 and the parties have made a *bona fide* attempt to resolve the dispute by mediation.'

The arbitrator could be faced with arguments that he lacks jurisdiction because there has not yet been a bona fide attempt to resolve the dispute by mediation. Again, the applicable mediation rules make provision for the mediation process to terminate automatically on the expiration of 42 days after the mediation commences or whenever the mediator feels that the possibility of a settlement has been exhausted. But as Lewis notes[26], unlike under the Government Form of Contract, where mediation is possible throughout the course of the construction works, under the MTRC Form of Contract, unless the parties agree otherwise, mediation is only possible after substantial completion of the works or following an earlier termination.

In a concluding article[27], a helpful update of recent developments is provided in regard to the use of the different contract conditions on the various airport related projects. First, it is indicated[28] that the problem of a bona fide attempt to mediate has now been dealt with in the MTRC documentation. A definition of bona fide is included:

> '... a bona fide attempt should be deemed to have been made provided the following minimum steps have been taken:
> (i)   A mediator has been appointed pursuant to the MTRC Mediation Rules; and
> (ii) The party serving a request for mediation has attended at least one meeting of the mediator (whether with or without the presence of the other party).'

Second, mediation notices no longer need to be served within 30 days of the completion of the works. However, the exclusion of mediation during the course of the works has not been removed.

In 1994 the Airport Authority introduced a four stage settlement procedure which incorporated for the first time in Hong Kong a dispute review panel as part of the dispute resolution process. Under clause 75 of the Provisional Airport Authority General Conditions of Contract (Civils/Building), a dispute notice is served by the aggrieved party on the other (clause 75.2). Disputes are dealt with by the project manager in the first instance (clause 75.3) within 30 days of notification, but may be subject to review by the project director (clause 75.5). If the project director fails to reach a decision within 60 days of referral of the dispute or either party is dissatisfied with the project director's decision, the dispute may be referred to the dispute review panel within 21 days of the expiry of the 60 day period or the date of the decision. Under clause 75.6, following a decision of the project director or a failure to reach a decision within the requisite period, disputes may be referred to the dispute review panel (clause 75.6 (a)) and/or arbitration (clause 75.6 (b)). Ordinarily no dispute can be referred to the dispute review panel, except with the consent of employer and contractor, after the completion certificate has been issued (clause 75.6(b)(i)). The use of the dispute review panel is an option in addition to arbitration. Any party still aggrieved can refer the decision of the dispute review panel to arbitration within 90 days of the date when made. The general powers of the dispute review panel are set out in clause 75.12, which says:

> '... the Dispute Review Panel and the arbitrator shall have full power to direct such valuation as may, in their or his opinion, be desirable in order to determine the rights of the parties and to ascertain and award any sum which ought to have been subject of or included in any certificate and to open up, review and revise any notice, withholding of permission or consent, certificate, instruction, request or decision of the Project Director or the Project Manager relating to other Dispute. The arbitrator shall have full power to order the rectification of the Contract ...'

Alongside the conditions of contract must be read the dispute review procedure[29], which consists of 12 pages setting out the detailed procedures. The procedure details arrangements to agree issues and appoint the panel (articles 2.2–2.5 inclusive) which can be one or three members depending on the dispute. Article 3 sets out the general duties of the panel with the procedure detailed in article 4. The panel has 'the widest discretion permitted by law to determine how to proceed ...' (article 4.1). Ordinarily only officers and employees will be present at meetings convened by the panel, but article 4.5 does allow limited use of outside consultants including lawyers. The panel has wide powers, which are set out in article 5 with the procedures relating to panel decisions set out in article 6. Awards of costs are discussed in article 7. A panel member cannot subsequently serve as arbitrator (article 8). All submission documents and decisions are ordinarily admissible in subsequent proceedings (article 11), although panel members are not compellable as witnesses. The procedure is generally overseen by a convenor whose role is defined in article 12 and who appoints the panel following the parties' representations. Usually the convenor will not be on the panel (or the chairman) (article 12.8) nor an arbitrator in subsequent proceedings (article 12.9) but interestingly, if the parties so agree, he can act as mediator under articles 12.10 and 14.

Although a little dated, a useful summary of organisations offering ADR services both in the UK and overseas can be found in the Construction Industry Council's booklet, *Dispute Resolution*[30].

Chapter 4
Mediation and Conciliation

Definition

Mediation and conciliation have been introduced briefly and a broad definition provided. In this chapter we look in some detail at what they seek to achieve and review their definitions because there is still much confusion in the use of these terms. The distinction is not purely academic; it is important because of the need to understand what process one is entering. The definition in itself is unimportant if one understands the process in which one is to become involved. However, it is often a matter of the parties' perception and if they attach meaning to words such as mediation and conciliation, then their definition is important.

The Construction Industry Council (CIC) publication entitled *Dispute Resolution*[1] recognises that different organisations use the terms 'conciliation' and 'mediation' to describe the same or different processes. For the purposes of the booklet it adopts the following definition in respect of both mediation and conciliation:

> 'It is where a neutral third party assists the parties in dispute to arrive at a settlement. In some schemes the third party is empowered to make a recommendation or impose a settlement subject to certain conditions which are virtually indistinguishable from Adjudication.'

This is not that helpful in that it gives an amorphous rather than a specific definition. Nevertheless, the CIC view, that different organisations use the terminology differently, is supported by references to the Academy of Experts (formerly the British Academy of Experts), the Centre for Disputes Resolution (CEDR) and the Chartered Institute of Arbitrators (CIArb). The Academy of Experts describes its conciliation/mediation service as:

> 'a process whereby a dispute between two or more persons or companies is resolved by remitting the dispute to a private hearing before an independent neutral third party (the Mediator) whose role is to assist the parties to reach a mutually satisfactory solution to the matters in dispute.'

The CIC and the Academy of Experts make no distinction between conciliation and mediation. The CEDR and CIArb distinguish conciliation from mediation. The CEDR sees conciliation as an informal attempt at bringing parties together to resolve a dispute and which subsequently may lead to mediation or other means of settlement. Mediation on the other hand is the slightly more formalised process of assisting the parties to negotiate a settlement. It seems that the CEDR approach is to treat conciliation as a first stage attempt to settle or as a prelude to other settlement processes, distinguishing itself because of its informality. The distinction made by the CIArb is more clear cut in that conciliation is defined as:

> '...a process whereby a conciliator investigates the facts of the case, attempts to reconcile the opposing contentions of the parties and prompts them to formulate their own proposals for settlement of the case by indicating the strong and weak points of their arguments and the possible consequences of failure to settle. The conciliator will not usually make a recommendation of his/her own for settlement of the dispute, however; he/she acts as a catalyst for settlement by the parties themselves.'

Key words in this definition are 'The conciliator will not usually make a recommendation of his/her own for settlement...'. This provides the clearest distinction between conciliation and mediation.

This definition is at slight variance to that contained within the CIArb's Consumer Dispute Resolution Scheme, which is only applicable to disputes concerning goods and services and not to those for work and materials.

Brown and Marriott[2] define mediation as:

> '...a process by which disputing parties engage the assistance of a neutral third party to act as mediator – a facilitating intermediary – who has no authority to make binding decisions...'

This bears little if any difference from the CIArb's definition of conciliation. They go on to say that conciliation is a term sometimes used interchangeably with mediation and sometimes to distinguish itself from the process of mediation. The glossary of terms adopted by the National Joint Consultative Committee for Building[3] provides the following:

> 'Conciliation maybe used to describe mediation ... or may be used to describe a process whereby the neutral person facilitates settlement but does not set out his or her views on the issues if settlement is not achieved.'

By contrast, although not an obvious definition, mediation is described as:

> '...a type of ADR process where the neutral mediator assists the parties to resolve their dispute by facilitating communication and helping them to evaluate the strengths and weaknesses of their cases...'

The particular conciliation procedures offered by the Institution of Civil Engineers (ICE), the International Chamber of Commerce (ICC) and the

United Nations Commission on International Trade Law (UNICITRAL) Conciliation Rules are dealt with later in this chapter.

Other definitions are adopted in other parts of the world. These are analysed in the Fédération Internationale des Ingénieurs – Conseils (FIDIC) report, *Amicable Settlement of Construction Disputes*, which shows that the words are used in a variety of ways and that perceptions of conciliation and mediation differ. Brown and Marriott concur with the CIArb in that 'conciliation' is sometimes used interchangeably with 'mediation', but they add that it is sometimes seen as a facilitative approach as compared with a more interventionist approach.

Perhaps, with so much nit-picking it is not surprising that others[4] have treated conciliation and mediation as the same thing. After all, how important is any distinction between conciliation and mediation? It may at first appear that any distinction is really only of academic interest. However, it may go beyond that and have practical significance. Universally accepted definitions that are clear and unambiguous do not exist. It is not absolutely essential that we have clear and distinct definitions for these terms, although it would be desirable for them to be defined.

In short, there are a variety of processes that are described as conciliation and or mediation and potential ADR users need to be aware of them. The prospective user must understand what process (or service) is being offered and consequently what can be expected from its deployment. Currently conciliation and mediation mean different things to different people and in some instances are interchangeable. Therefore, until firm definitions are established one must determine what each term is describing and therefore what one can expect from the process.

A distinction between conciliation and mediation will never be easy to evince. The dictionary is not particularly helpful in this regard. Mediation is best used as a generic description, with conciliation reserved for a particular aspect of mediation, in the same way as negotiation. Should other approaches emerge they can then be ascribed other terms. At present this lack of clarity does not assist the advocates of ADR as in the attempt to avoid, inter alia, the mystification of arbitration and litigation, they may have inadvertently introduced another form of confusion. The proliferation of organisations wishing to be involved in ADR has not helped in this regard.

Mediation for the purposes of this book is defined as a non-judicial process that engages an independent third party (the mediator) as an intermediary, to assist the parties to bring about a resolution of the dispute(s) that exist(s) between them.

Approaches to mediation

There are generally considered to be two basic approaches to mediation: the facilitative approach and the evaluative approach. However, there are also versions of each in terms of process and also in terms of the mediator's role.

The facilitative approach, which is also referred to as the interest-based approach, is generally thought to be the purest form of mediation. The mediator is interposed between the parties to explore their positions, to provide a means of communication, to enhance their common interests and to provide an ambience that is conducive to the parties finding their own solution to their dispute. Generally, the mediator would not express an opinion or propose a settlement in this form of mediation. However, views also differ on this point.

By contrast the evaluative approach, which is also referred to as the rights-based approach, focuses on the respective rights of the parties in dispute. In this approach the mediator attempts to evaluate, with or without expert help, the strengths and weaknesses of each party's case and indicates a view. The purpose of this is to influence the parties to modify their positions so that the dispute may be resolved.

There is a range of views as to what the mediator should be doing and which is the right way to conduct the mediation. Mediation is the same as negotiation but the process is assisted by a neutral third party, who helps them follow good practice[5]. Here the mediator is seen partly as a teacher who shows the parties how to resolve their dispute. Naughton[6] believes the mediator to be a catalyst who should not at any stage express an opinion. It is unclear whether this refers to every mediation but if it does, only the facilitative form could exist, as the view is totally incompatible with the evaluative approach. Dispute resolution is sometimes viewed as interest-based bargaining in that it seeks to focus on satisfying as many as possible of the important needs and interests of the parties. It is not about winning, nor is it about compromise[7]. Writers who seem to prefer the facilitative approach as the way forward make the above points. Furthermore, as facilitator the mediator does not become pro-active by proposing or attempting to impose a solution, but rather promotes a realistic understanding by each party of the other's interest. Elsewhere, as illustrated by the drafters of the Hong Kong Mediation Rules for Public Works Contracts, the evaluative approach is preferred, at least initially, no doubt in the belief that the rights of the parties are the proper starting point for any settlement.

The facilitative approach has much in common with principled negotiations[8]. The use of principled negotiations involves deciding issues on the merits by some objective standard rather than resorting to positional bargaining. The latter approach encourages each side to take a position and stubbornly hold to it, rather than focus on their underlying concerns and needs. The four fundamentals for principled negotiations were identified as:

(1)   People – separate the people from the problem
(2)   Interests – focus on interests, not positions
(3)   Options – generate a variety of possibilities before deciding what to do
(4)   Criteria – insist that the result be based on some objective standard.

In either the facilitative or evaluative approach the mediator may be interventionist. It is the nature of the intervention that is the distinguishing factor.

In either aproach it would appear possible for the mediator to offer solutions, rather than relying on the parties to find their own. In the facilitative approach the mediator bases the proposal on the best interests of the parties, whereas in the evaluative approach the proposal would be based on an assessment of the rights of the parties. Notwithstanding this, it is preferable for the mediator not to offer a solution in the facilitative approach because it is arguable that only the parties can possibly know what is in their best interest. Therefore the mediator should only bring them to a point where they determine this for themselves. If the evaluative approach is adopted it would seem that the mediator may not only propose a solution but may work aggressively towards getting the parties to accept the proposal. This latter approach is one adopted by Feinberg[9] who believes that to come to a quick decision on the settlement position and then apply pressure to the parties to accept it, is the most productive way to proceed. An approach of this kind would be philosophically incompatible with the facilitative approach.

The Construction Disputes Resolution Group (CDRG) mediation service sets out in its guidelines that the mediator has the discretion to adopt any procedure that suits the parties and may at any stage make proposals for the settlement of the dispute. This provides a high degree of flexibility in that the approach adopted can be either facilitative or evaluative and may use any of the available processes.

The organisation, or an individual, seeking mediation may be better disposed to one or other approach. There may for instance be a philosophical problem with adopting the facilitative approach because it is not based on the agreement entered into. This could be the situation with a public body where accountability needs to be overt and obvious. On the other hand, others may not be inclined to use the evaluative approach because its fundamental philosophy is not dissimilar to arbitration, litigation or for that matter adjudication; that is, a third party assesses the dispute to identify the rights of the parties. In other ways the processes are of course quite different.

Because both facilitative and evaluative mediations exist, it is possible that the parties will not wish to adopt the same basic approach. This is problematical, regardless of whether it is established pre-contract or post contract. Arguably, it is better to resolve this issue before entering into the contract, otherwise the chance of mediation working would appear to be substantially reduced. One also has to recognise that this may be an inherent flaw in mediation but then anything that requires voluntary involvement could be said to be defective. As with most things there is no one right way to proceed but a number of methods are available which may or may not succeed, depending on what one seeks to achieve and the other variable factors appertaining to the dispute, not least the parties themselves.

The number of organisations, both within the UK and overseas, which offer ADR services has grown significantly since 1990 following a lead from the USA. These organisations frequently offer a choice of approaches and have their own view as to what constitutes mediation.

Mediation is always voluntary because it cannot take place without agreement. However, the stage at which this agreement is reached can be an important factor. For instance, it may be agreed to incorporate a clause into the contract that requires the parties, should a dispute arise, first to follow an ADR procedure. Whether an ADR clause can oust or rather suspend the jurisdiction of the courts is problematic, but any direct approach to the courts that sought to circumvent these prerequisites would generally be met with a stay of such proceedings.

Where an ADR clause is incorporated in the contract it will not necessarily prescribe mediation as the process to be used should a dispute arise. Nevertheless, mediation is generally made available as an option within the clause provisions. In order that the clause has a reasonable chance to operate and is not subject to sterile wrangles, it is better to specify to which particular ADR organisation the dispute will be referred. Thus the approach to ADR, although not absolutely determined, is in part prescribed. Even where the provisions or adoption of a particular service spell out a particular process, this may be modified or ignored if the parties agree. The parties can opt out of the process in the same way as they opted in – by agreement.

Where an ADR clause is not written into the contract, there is no obligation at present to use ADR before proceeding to arbitration or litigation. However, there may be sound reasons, many of which are set out in this book, for considering this alternative approach. Reaching agreement after a problem has arisen is always fraught with difficulty. This can be witnessed in arbitration, with the consequent adoption of an appointing body and rules of procedure being incorporated into the agreement prior to commencing work. Similarly, ADR is most likely to be started where it has been incorporated into the contract provisions. Otherwise, to propose ADR is seen as negative from a claimant and brings the taunt 'litigate or shut up'.

The mediator

The choice of mediator is a vital part of mediation. How one tackles this issue may influence not only the basic approach to mediation but also the detailed procedures to be followed. Perhaps more importantly it will influence the whole environment within which a settlement is sought.

An informed user of mediation may have a view of what is required, with a clear preference for one or other approach. The approach chosen could then influence the selection of the mediator, choosing one who has the appropriate skills for the favoured process. However, the view of one party on this is unlikely to hold sway and both parties would need to be of the same view before this is likely to occur.

At present the use of mediation in the construction industry within the UK is fairly limited and it would seem that selection of the type of mediation preceding the appointment of the mediator would be unusual. More likely, parties seeking an alternative to litigation and/or arbitration would approach one of the organisations offering ADR services, seeking the nomination of a mediator. As most of these organisations offer mediation procedures, this would frequently lead to the adoption of that organisation's procedure.

Ideally parties should determine which basic approach to mediation they are comfortable with and find acceptable, before choosing the mediator. This is best incorporated into the contract, together with a mechanism to appoint the mediator, i.e. which organisation is to make such an appointment. The selection of this organisation also requires care and will generally be one perceived as a brand leader. It must be sensitive to the wishes of the parties and able to identify a mediator who is sympathetic and has the skills to fulfil the particular requirements of the selected mediation approach.

Mediators should conduct themselves in a 'proper' fashion if ADR is to become respected and trusted. The Academy of Experts mediation summary provides a useful Code of Conduct for Mediators:

- A Mediator should not accept an appointment where there is actual, potential or apparent conflict of interest between him and any one of the parties. However, if the parties are aware of the potential conflict, they may endorse his appointment anyway.
- Mediators cannot work on contingency fees or indorse arrangements whereby the loser pays.
- A Mediator has a duty to comply with the Academy's Guidelines.
- A Mediator must maintain neutrality.
- A Mediator must refuse to act as a witness, advocate or adviser in any subsequent litigation relating to the dispute.
- Notwithstanding the exclusion of personal liability, the Mediator must have suitable Professional Indemnity Insurance.
- No Mediator should publicise his services in a way which could be construed as in bad taste. He should not indulge in inaccurate or misleading publicity.
- At all times the Mediator must comply with the Academy Guidelines and Code of Conduct.

Functions of a mediator

The functions of a mediator are partly affected by whether the facilitative approach or evaluative approach is to be followed. The facilitative approach primarily aims to recognise the business interests of the parties with negotiation of a commercial settlement. In contrast, the evaluative approach is based on the assessment of the contractual rights of the parties and persuasion by the mediator towards a settlement based around this appraisal.

As facilitator the mediator will:

- create a conducive environment that ensures an appropriate level of confidentiality

- familiarise the parties as to the mediation process
- establish protocols for conducting the mediation
- assist in bringing the parties together and encouraging face-to-face discussions
- help the parties understand each other's viewpoint and examine mutual concerns
- help identify common ground, isolating the really contentious issues
- prioritise the issues and focus on a plan as to how each can be dealt with
- get the parties to think actively about a settlement and the consequences of having to adopt an alternative means of settlement
- explore possible solutions and assist the parties to formulate settlement proposals
- provide or procure relevant problem solving skills
- act as a sounding board for any ideas the parties may have as to a settlement
- keep the parties talking
- restructure the parties' communications so as to remove emotive words and improve understanding
- assist in preparing a draft settlement based on the parties' agreed position
- seek to bring the parties into a binding settlement.

As evaluator the mediator works in a similar way, but there are some distinct functions (shown in italic) which are discrete to the evaluative approach:

- *analyse the problem and provide or procure the relevant analytical skills*
- *establish an evaluation of the dispute*
- help identify common ground, isolating the really contentious issues *and expressing an evaluation of the dispute*
- get the parties to think actively about a settling *along the lines of this evaluation* and the consequences of having to adopt an alternative means of settlement
- *negotiate or persuade the parties to formulate settlement proposals based on the evaluation.*

Skills of a mediator

Whether the mediator is an evaluator or facilitator there are a number of common functions and hence some common skills are necessary. However, some dissimilar functions exist and the skills required for these are very different. Brown and Marriott distinguish between the attributes required of all mediators and the skills required. They identify the following attributes as necessary: understanding, judgement, intuition, creativity, trustworthiness, authority, empathy, constructiveness, flexibility and independence.

This list is part of what makes up the personality of the mediator and could be extended. Important attributes that seem to be missing from this list are: objectivity, assertiveness, patience and tenacity. Assertiveness is an interesting and problematic attribute, in that it is required at varying levels within the mediation process and can alienate the parties if seen as judgmental.

Certain of the skills required are inextricably entwined with the attributes. Certain high level skills require particular attributes; for example, a negotiator needs the majority of the above attributes. So does the skill of conflict management.

The attribute of independence – the ability to work autonomously without support or feedback and to maintain a neutral and independent stance – is stressed by all advocates of ADR. It seems an obvious attribute but is not axiomatic, because what is independence? One can be independent of mind and indeed action, without necessarily being independent of those associated to the dispute. The Academy of Experts refers to the mediator as an 'independent neutral third party', stressing that the mediator has to be independent of mind and action and free of association with the disputants. With regard to 'neutral', if this is taken literally it means that the evaluative approach cannot be adopted because neutrality means not taking part on either side.

It is wise to select a mediator who is independently minded and not associated with the disputants, but that does not mean that bias will not be present. Bias of some description is virtually impossible to avoid. Everyone has prejudices, reads newspapers, watches television and talks to other people. The irony is that one may select an independent mediator who is less helpful in resolving the dispute than a non-independent mediator. Nevertheless, it is impracticable to accept the latter because human nature would assume that this must work against the interests of one party.

The main difference between the skills required for the facilitative and evaluative approaches can be summarised as follows. In the facilitative approach, negotiating skills dominate. It is not essential, and may be disadvantageous, to possess expert knowledge and skills pertaining to the area of the dispute. There are differing opinions as to the value of expert knowledge in this process. It can be useful if used correctly in the exploration of solutions, but it can mean the mediator unknowingly adopts a position. In the evaluative approach it is essential that such knowledge and skills exist because without them a proper evaluation cannot be made. These skills are likely to include both legal and technical competencies. In addition, a higher level of assertiveness will be required to persuade the parties to accept a solution based around the evaluation.

Most writers on mediation stress the importance of selecting the right person to perform the role of mediator. The person selected will need to have the requisite attributes and skills but trying to identify such a person is far from easy. Because there is a need for them to be independent third parties, they need to be at a distance from those requiring their services. This will often mean that potential mediators, who are currently a small 'pool', are unknown to the disputants. Even to consider the recommendation of one of the consultants on the project raises an innate wariness.

It is not difficult to see, therefore, why those who find themselves in dispute and wish to pursue ADR may contact an ADR organisation for general assistance and the nomination of a mediator. The responsibility that these organisations have for the success of mediation is weighty. Training is vital but it would be wrong to assume that anyone can be trained to become an effective mediator. Without the prerequisite attributes no amount of training will produce what is needed. Eventually the best mediators will emerge and develop a reputation. Whether they will be available to resolve a dispute when required and at a sensible price (shades of arbitration) is another matter, again raising the spectre of finding yet another means for the resolution of disputes.

Mediation agreement

The mediation agreement needs to be clear so that both parties know what to expect from the mediation. Paragraph 3 of the JCT Practice Note No. 28 provides a list of matters that may be covered in a mediation agreement. These include:

- a clear and precise statement of the issues in dispute
- a declaration by the parties that they wish to use a mediator to facilitate settlement
- a period during which the mediation is to take place
- the name and qualifications of the mediator
- the mediation is to be conducted on a confidential without prejudice basis unless the parties agree otherwise and the mediator will not be called to give evidence in any subsequent proceedings
- a date for the meeting with the mediator to discuss format
- how costs arising out of the mediation are to be dealt with
- if the mediation results in a settlement of some or all of the issues, the parties will execute a binding agreement setting out the terms of settlement.

Usefully the Practice Note concludes with some draft agreements and these include an example of a mediation agreement.

Mediation procedure

There is no set procedure for mediation. ADR organisations often have guidelines or codes of practice but seldom impose a tight procedure to be followed. For example, as referred to earlier, the CDRG in item 4 of their guidelines state that:

'The procedure to be adopted will be at the discretion of the mediator who will give due consideration to the wishes of the parties.'

Item 6 of the guidelines says:

'At the meeting the mediator will assist the parties to reach an amicable settlement of their dispute and will conduct the proceedings in such manner as he considers appropriate.'

Much will depend on the parties, the mediator and the basic approach to be adopted. Notwithstanding this inherent flexibility, which is seen as a fundamental strength and virtue of mediation, some procedural good practice has been developed. Aspects of this include:

- Before any involvement by the mediator, the parties will submit to the mediator a brief statement that summarises their view of what is in dispute.
- The mediator will ensure that both parties are provided with a copy of this statement.
- The parties informally present their cases to the mediator in the presence of the other party (avoid over formalising this part of the process otherwise rigidity follows and a mirror image of other dispute forums may be generated).
- The mediator will discuss the issues raised by the submissions and presentation of the cases in an attempt to isolate the key points; this may be done with both parties present or with each party in turn – referred to as 'shuttle diplomacy'.
- The mediator will discuss in confidence the strengths and weaknesses of the parties' respective cases and will get them to focus on what is in their best interests.
- The mediator will extract ideas, in confidence, for resolving the key points of the dispute and will formulate them into outline proposals; these can then be used to identify any common ground.
- The mediator can then assist the parties to put their proposals for settlement to the other party at a joint meeting, if private meetings have been used prior to this stage.
- The mediator will then seek to close any gap that exists between their respective positions.
- The mediator will draw up a draft agreement based on these proposals, negotiate any outstanding points and secure a binding settlement that can be enforced in law.

In other processes the mediator may offer a solution or settlement proposal if negotiations founder, but this depends on what the parties want from the mediation. A truly facilitative mediation would not involve this process.

Some exponents of mediation believe that the discussion following the initial presentations should always be done in private. These sessions are referred to as caucuses. The term 'shuttle diplomacy' is used to describe these meetings and the mediator's role. Caucusing is favoured because the mediator can build up a clear view of the respective cases, private agendas and possible settlement positions, without each party disclosing to the other party their particular strategy towards settlement. It also allows the mediator to tease out weaker points.

Some believe that the secrecy of caucusing can lead to apprehension, which creates the wrong ambience for settlement, but a completely open forum also has drawbacks. A joint meeting makes the parties wary about what they say and may produce a sterile situation rather than the fertile one that is necessary for mediation to work successfully. Issues of this kind raise an array of ethical considerations and the introduction of codes of practice attempts to regulate these concerns.

An evaluative mediation involves some different procedures. However, there are common aspects, especially in the early stages:

- Prior to any involvement by the mediator, the parties will submit to the mediator a brief statement that summarises their view of what is in dispute.
- The mediator will ensure that both parties are provided with a copy of this statement.
- The parties informally present their cases to the mediator in the presence of the other party. (Over formalising this part of the proceedings should be avoided.)
- The mediator will discuss the issues raised by the submissions and presentation of the cases in an attempt to isolate the points on which an evaluation will be based. This may be done with both parties present or with each party in turn.
- The mediator will analyse the dispute against the background of the respective rights of the parties and will produce an evaluation of the dispute.
- The mediator will discuss in confidence the strengths and weaknesses of the parties' respective cases and will persuade them to accept the evaluation.
- Where the parties are unable to accept the evaluation, the mediator will seek to identify the settlement zone and formulate outline proposals for consideration.
- Agreement will be sought on the reformulated evaluation and the parties will be persuaded or even coerced by increasing assertiveness on the part of the mediator. (Feinberg's attitude is that the mediator should be very assertive, perhaps even aggressive in seeking settlement, once his evaluation has been made.)
- The mediator will draw up a draft agreement and secure a binding settlement that can be enforced in law.

Although the two basic approaches are different philosophically and procedurally, they are not mutually exclusive in all respects. It is not uncommon for them to be used consecutively in mediation, often with the facilitative approach following the evaluative approach where the latter has failed to find a solution. They are less likely to be used the other way round because one is more inclined to proceed to arbitration and or litigation once a facilitative mediation has failed to produce a settlement.

Whatever type of mediation is used, it is essential that the parties are represented by someone who has the authority to conclude an agreement. It is also desirable to set a time scale in which settlement is to be achieved as this focuses the mind and is good dispute management. For example, the CDRG set a limit of five days for mediation.

Guidelines for mediation

Various sets of guidelines for mediation have been produced and some of these are set out below. One of the most useful is from the Academy of Experts, provided under the following headings:

- *Appointment of a mediator*
  If the Academy is the appropriate body to oversee the mediation, a short outline of the dispute will be provided to the Academy to assist in the selection of the appropriate neutral. The appointed mediator will require the written acceptance of the parties before beginning the mediation. It is essential that the mediator disclose to the parties any interest he may have in the subject matter of the dispute or connections with any of the parties. The Academy emphasises the need for the mediator to maintain his integrity, to avoid any conduct which may call into question his impartiality and to be aware of any personal factors which might compromise his ability to be or appear fair.
- *Meeting of the parties with the mediator*
  Within not more than two weeks of his appointment, the mediator will arrange a meeting of the parties. In addition, he may request, at least seven days before the meeting, that each party provides him with a brief memorandum setting out the relevant facts and issues in dispute and their position on the questions raised. The mediator will circulate this memorandum to the other parties. In order to avoid the mediator being deluged with documents, the exchange of information will be controlled by the mediator. Under the Academy's guidelines the mediator may, with the agreement of the parties, make a site visit or carry out an inspection or seek legal or other technical advice.
- *Conduct of the meeting*
  The Academy emphasises that the parties should bring to the meeting all documents and information on which they may wish to rely. The mediator is master of his own procedures:

- he can combine joint and separate sessions with the parties, as he sees fit;
- as the situation demands, he decides which party should make the opening position statement.

Under the Academy's guidelines the proceedings are neither tape-recorded nor a transcript professionally taken by a stenographer. The parties may appear in person, be assisted by lawyers, or work out a flexible approach involving their own advocacy and that of their lawyers. In the absence of agreement being reached, the parties may request the mediator to make a report containing his findings on how the dispute should be settled. This equates broadly with a 'recommendation from the conciliator' under the ICE Conciliation Procedure 1994.

- *Termination of the mediation*
  As with certain other dispute resolution schemes, a mediator has a discretion to terminate a mediation at any time if he believes that the mediation will not be successful. Similarly, any party to the mediation may withdraw. At the conclusion of the mediation, whether successful or not, the mediator will return to the parties all documents provided to him and destroy any notes that he has taken.

- *Confidentiality*
  The Academy emphasises the private nature of mediation as a method of dispute resolution. The Academy states that mediation is a bona fide attempt to resolve disputes between the parties with the entire process consequently being without prejudice. To protect a mediator's position, anything stated or that emerges or that is disclosed during the mediation is not discoverable or admissible in any subsequent formal proceedings unless the documentation is of a nature that would not ordinarily be subject to legal privilege. To avoid any question of the mediator being called as a witness in later litigation or arbitration, it is stated in bold type:

  > 'No party may call the Mediator as a witness in any subsequent legal proceedings to give evidence concerning matters disclosed during mediation.'

- *Exclusion of liability*
  Unlike a judge or arbitrator who enjoys immunity from claims that might otherwise be brought as a result of dissatisfaction with the manner in which a court case or arbitration was conducted, a mediator probably (the point has not been litigated) owes a duty of care in negligence to each of the parties as well as a contractual obligation under the contract appointing him to provide his professional services. Therefore, it is essential for a mediator's peace of mind to ensure that all liability is

excluded for what he may or may omit to do during the course of the mediation. This is set out in the Academy's guidelines, with a further exclusion to the effect that the Academy is not responsible on an agency basis for the conduct of the mediation.

- *Costs*
  As under other similar schemes, the parties are responsible for their own costs and jointly liable for the costs of mediation.

- *Special Note*
  Although written in somewhat muted language, there is an important warning for all those parties who might consider mediation an option for their dispute in place of or in addition to conventional litigation or arbitration. If the parties' contract has a well-defined disputes procedure, the parties should agree how this is to be modified to accommodate ADR. The Academy also warns that it may be appropriate to take specialist legal advice. The contract may contemplate dispute resolution via the courts or arbitration. If so, a number of issues arise. First, the enthusiasm to use ADR must not lead to a disregard for the possible problems posed by the Limitation Act 1980 and associated legislation. In the case of a contract executed under hand, legal action must be commenced within 6 years of the breach of contract occurring. If the contract is executed under seal, as many construction contracts are, the limitation period is 12 years from the date of the breach of contract. If the claimant's case is based in negligence the limitation period is ordinarily 6 years from the date of damage occurring, but if the damage is of a hidden (latent) kind and could not ordinarily have been ascertained by the claimant during the primary limitation period of 6 years, the claimant has, by virtue of the Latent Damage Act 1986, a further 3 years in which to bring proceedings based on the original negligence, albeit subject to a limitation cut-off date of 15 years from the date when the cause of action originally accrued. It is therefore essential that any party embarking on mediation is conscious of problems that may be posed by the expiration of the limitation period. It is one thing to attempt mediation and to be unsuccessful, but quite another to embark on mediation only to find that on its failure there is insufficient time left to pursue the claim through the courts or arbitration.

Chartered Institute of Arbitrators

The Chartered Institute of Arbitrators (CIArb) offers alternative dispute resolution services, including conciliation, mediation and supervised settlement procedure (mini-trial). In 1990 the Institute issued a series of booklets outlining the services and providing guidelines for each of these dispute resolution processes.

The limited success of these various initiatives can be seen in the CIArb's failure to publish updated leaflets which, at the very least, state the CIArb's current address.

CIArb guidelines for conciliation and mediation

The CIArb views conciliation and mediation as fundamentally the same and provides a common set of guidelines. The principal difference is that the mediator would recommend a settlement, whereas the conciliator would not. The guidelines cover the following matters:

- application of guidelines
- procedure for commencement of proceedings
- appointment of conciliator or mediator
- conciliation/mediation procedure
- settlement of disputes
- termination of proceedings
- arbitration or litigation following proceedings
- costs.

Proceedings commence with an application on a one page standard form provided by the CIArb. Only brief details of the dispute are required for identification purposes at this stage and the parties must indicate whether they are seeking a conciliator or mediator.

The guidelines set out that the proceedings are confidential and privileged. This is important if the parties are being encouraged to disclose their respective positions, including what they consider to be their strengths and weaknesses. If the dispute is not settled in these proceedings it may subsequently proceed to arbitration or litigation but no one associated with the mediation or conciliation may be called on or compelled to give evidence about what happened at the conciliation or mediation. Furthermore, the conciliator and mediator cannot act in any capacity in subsequent proceedings relating to the same subject matter, unless the parties agree. In order to protect fully the parties' positions, should the matter proceed to litigation or arbitration, the confidentiality extends to a range of matters that cannot be subsequently raised: the views and admissions of the parties and anything related to a proposal or abortive draft settlements. In practice the parties' position can never be quite the same subsequently because knowledge will impact in a variety of ways and likely shape the course of subsequent events.

It will be usual for the CIArb to appoint a sole conciliator or mediator but the parties may each appoint one and the Institute be required to appoint a third. Although the parties may make representations to the CIArb as to who it should be or the qualifications to be held, the CIArb can make the appointment as they see fit. Proceedings should start immediately after the appointment and can take any form appropriate to the dispute. The conciliator or mediator will fix time limits within which the parties are required to work. The proceedings should be concluded no later than three months after commencement unless otherwise agreed, although it is clearly in the parties' interests to conclude matters earlier if practicable. Although the conciliator or mediator has significant discretion in how to proceed, the guidelines require that account be taken of five matters:

- circumstances of the case
- business relationship of the parties
- parties' wants
- speed and cost
- whether documents only will suffice.

The powers of the conciliator or mediator are wide and include the ability to interview the parties separately and to require them to provide security for costs. Where separate interviews take place, the disclosure of information by one party to the conciliator or mediator can only be subsequently revealed to the other party for comment if prior consent is obtained. This protection is to encourage the parties to speak openly, so that the conciliator or mediator can appraise the issues that underlie the dispute as well as the dispute itself. Such knowledge will help the conciliator or mediator find a way forward although the restriction of not putting it to the other party can be a drawback. However, if the conciliator believes the disclosure of such information will expedite a speedy settlement an agreement to disclose it should be sought.

Early settlement of any dispute is an objective and therefore the parties are to be encouraged constantly to seek a solution and to offer their own proposals. The conciliator or mediator may offer a preliminary view at any stage. Proposals are required from the conciliator within three weeks from the conclusion of 'an informal hearing, a site visit, or receipt by the conciliator or mediator of legal or expert opinion'. By contrast, the mediator *may* formulate proposals but is not required to do so. Where a proposal is provided the parties may comment on it and the conciliator or mediator may revise the proposal accordingly. When a settlement is achieved, a settlement agreement should be drawn up to give it contractual effect and make it enforceable in law. It is unimportant who draws up the agreement so long as it is legally binding and does not in any way change the nature of the settlement.

Because reference to conciliation or mediation under this scheme is voluntary, either party can withdraw at any time prior to a binding settlement being achieved. If the matter under dispute is referred to litigation or arbitration this is tantamount to the party withdrawing from the conciliation or mediation. Such action is threatening and not conducive to conciliation or mediation and therefore proceedings are terminated under the scheme. If the conciliation seems likely to deal satisfactorily with only part of the dispute, one party should not pre-empt this by referring the outstanding matter to litigation or arbitration before the proceedings are concluded. To do so would bring proceedings to a premature end. Proceedings can be terminated if the conciliator or mediator considers it futile to continue.

The costs of conciliation or mediation proceedings are borne equally by the parties and section 8 of the procedure sets out what items of expenditure are included as costs. In addition to making an order for security of costs, the conciliator or mediator can order the deposit of costs and may suspend or terminate proceedings until there is compliance with an order. What constitutes acceptable security for costs is for the conciliator or mediator to determine. The procedure does not mention where the costs are to be held and whether any interest on the deposit should accrue.

Progressing the mediation

Before the formal mediation sessions commence, the seating arrangements for the parties need to be addressed. The opposing camps must not be seated in such a way as to create in their minds the feeling that a particular party is being advantaged. Ideally the representatives of the various parties should be equidistant from the mediator at the opening session so that eye contact can be engaged with any necessary person. For that reason a round table is probably best avoided. Depending on the number of parties either a rectangle or an H configuration is better. Both arrangements place the parties' representatives equidistant from the mediator, allowing him to engage either side in dialogue or listen to a point without antagonising the other.

Once the mediator has convened the parties at the first mediation session, he will make some opening remarks indicating to the parties what he hopes to achieve. A good mediator should put the parties at their ease, not take sides, and should encourage the parties to consider issues rather than dwell on personalities. He will confirm that mediation is a non-binding form of dispute resolution, which will impose no solutions on them. If the parties have submitted written position papers in advance to him, he will indicate to them that he has read them and will then endeavour to ensure that a co-operative negotiating strategy is adopted rather than an aggressive or hectoring tone. Next he will ask each side to make a short opening statement, which will often last no more than 10–15 minutes, during which they explain their position.

If position papers are submitted in advance to the mediator, they should be shorter rather than longer; perhaps not to exceed 20 pages of double spaced A4 typescript. Although the precise format of position papers will depend on the nature of the dispute (or might be dispensed with altogether) they might include the following:

- introductory remarks and a positive indication that it is the client's wish to work towards settlement
- a résumé of the facts of the case as seen by the client but highlighting any

agreements or disagreements that are believed to exist in regard to the particular facts
• an analysis of liability and quantum.

Position papers should be meticulously prepared in order to strike just the right note, contain the appropriate information (without being 'tell all' documents) and have the correct balance of factual and technical content. Flowcharts, diagrams, photographs and plans may all be useful for inclusion.

The subsequent oral statements should be as neutral as possible, setting out the issues and drawing attention to the interpretations and conclusions of each party. It is useful for these statements to have been prepared and practised by lawyer and client in advance of the formal mediation session, particularly if client representatives of sufficient authority, experience and detachment from the dispute can be used as independent critics.

Those involved in preparing and making the oral submissions must be realistic and honest. Although a litigator may present a case in the most favourable way by emphasising strengths and ignoring weaknesses, which is legitimate under the adversarial system, this is unacceptable in mediation if carried doggedly into the private caucus sessions. As with the position papers, the oral submissions should avoid specific settlement figures, should not emphasise what are considered to be deal breakers and should avoid at all costs emotive language. It would be futile to commence any mediation session with the statement, 'I'll settle for £50,000 and not a penny less'. The other party will simply be thinking up all the reasons possible why there is no entitlement to £50,000, instead of listening to the facts of the case as presented by the first party.

In his handling of the mediation sessions, the mediator must remain aware throughout that the parties do not wish to be cajoled or coaxed into a compromise settlement and that mediation is not simply about 'splitting the difference'. It is all about trying to achieve 'win win' solutions. Methods of case presentation are flexible. It is legitimate for the parties to rely on all the presentational aids that they can muster, including visual ones such as photographs, site diaries, clerk of works records, plans, specifications, overhead projector slides which highlight key documents and elements of expert reports on which a party may wish to rely, flow charts, and if the case involves defective work, cores taken and other materials removed from site. If the claim relates to delay and disruption and critical path analysis, this can be presented either as computer graphics or as simple overheads. Any necessary *locus in quo* report can be made by video.

There is no set period for mediation that will achieve success. Mediation is a process of trying to respond to the parties' personal requirements while, at the same time, reducing their adherence to positions that are not justified. The common ground and shared values have to be identified and confirmed and distinguished from weak bargaining. This strikes a positive note, may immediately reduce the areas in dispute and provides each of the parties with a feeling that the process is beneficial as they recognise the common ground. Some people suggest that to achieve this effectively it is useful to have some experience of the customs and practice of the industry involved.

Two examples of how mediation can work and identify a hidden try to settlement are given here.

Mr and Mrs X and Mr and Mrs Y were owners of adjoining properties. A large bushy hedge marked the boundary at the rear of both properties. Mr and Mrs X went on holiday and found on their return that the hedge had been removed by Mr and Mrs Y and a fence erected. Mr and Mrs X asked Mr and Mrs Y to remove the fence, which they failed to do. Mr and Mrs X went to their solicitors who informed Mr and Mrs Y by letter before action that if the fence were not removed to the actual line of the boundary within 7 days, legal proceedings would commence in the County Court. The fence remained and the proceedings started. Damages and/or an injunction and a declaration as to the line of the boundary were sought. Those proceedings limped on for 6 years until mediation was recommended. During the process the mediator asked Mr and Mrs X what they wished to achieve. Mrs X said that was obvious – they wanted an apology! The mediation was satisfactorily resolved in 30 minutes.

The need to recognise the hidden agenda of a particular party emerges from the following example.

A roofer sustained terrible injuries when he fell into the well of a building. Liability was not an issue from the defendant's solicitors/insurers; the only arguments related to quantum of damage. The plaintiff argued that he required £200,000 to adapt his home to meet his requirements. The insurance company was not prepared to offer more than £100,000. Instead of legal proceedings mediation was used during which it transpired that, for *policy* reasons, the insurers could not offer £200,000 under this particular head of claim. However, the insurers were happy to provide the plaintiff with a further £100,000 on the basis that it was spread across other heads of claim and thereby lost.

Types of dispute suited to mediation

Mediation has been used to settle disputes in a wide range of different activities. Initially developed to resolve labour disputes, it has subsequently been used in community and neighbourhood disputes, public policy matters, social conflict, family matters and health care as well as commercial disputes. The use of mediation in the construction industry is a more recent development.

The proponents of mediation point to the savings that can be made as compared with either arbitration or litigation. Louis Selig[10] is quoted as saying that analysis of past results shows that every $1 spent on mediation would save $10. Quick[11] makes several references to the claims made by ADR and how the high cost of arbitration and litigation has encouraged a worldwide development of ADR. He also refers to a report[12] by the Australian Federation of Construction Contractors which provides information related to the low cost of ADR and the high level of success but questions the reliability of these and other claims because empirical proof is difficult to secure. In the light of the claims made for ADR, and notwithstanding the apparent reservations, it is not surprising that some disputants are prepared to try ADR. But whether it is equally applicable in all fields, and construction in particular, remains an open question.

In areas where disputes and conflict arise and where there is no clear framework or reference point, mediation has something to offer; for example, in labour relations and family matters. Both parties in a majority of these situations have their own perception of what they want, although it is not necessarily based on any solid foundation or pre-existing agreement. The facilitative approach has much to offer in trying to bring the parties to common ground by exploring their interests.

Commercial disputes arising out of a contract are quite different and it is this type of dispute that predominates in the construction industry. Such disputes do not lend themselves so well to the facilitative approach because there is often a contract that sets out their respective rights and obligations, to which someone can point. With partnering in vogue and high legal and other bills for everyone when confronts fail, ADR should have something to offer. Therefore, if a contractual agreement exists why not use it for settling the dispute? If one cannot afford the costs of arbitration and/or litigation, an evaluative mediation would seem the way forward. To act otherwise suggests that one prefers to negotiate a new agreement rather than resolve the matter based on agreed rights. A cynic might suggest that the parties seldom know what they have agreed when they enter a contract and consequently it is better to negotiate what that contract might be once the relevant facts are known. In other words, negotiate after the event. But if a contract exists and the risk is apportioned and priced, to mediate other than on the rights arising from the contract seems rather strange. The only reason that ADR exists is because the systems for determining and enforcing those rights are ineffective in terms of cost and time.

Construction disputes are suitable for mediation by the evaluative approach; mediation by the facilitative approach is less attractive. Arbitration and litigation will remain inevitable in certain circumstances and if these processes were streamlined they could still have much to offer, with the inherent benefit that the decision is enforceable and the procedure played out under specific rules. Some writers[13] believe that ADR is well suited for construction contracts because of the challenging work required in construction to be completed on time and to budget. But is mediation suitable for all construction disputes? This is difficult to answer. For example, is mediation suitable if those directly involved with the project can find an accommodation? 'There is no question that certain accommodations reached between the engineer and contractor make sense and are necessary to promote smooth project performance'[14].

Is this an attempt to resolve a dispute by informal means? Is this an attempt to avoid a dispute arising? Does this mean the promoter's (client's) contractual position is being compromised, albeit for what is considered the best longer-term interests? In reality, it can be either dispute avoidance or dispute resolution and may or may not compromise the promoter. If it were dispute resolution then this would be one way of pre-empting a reference to more 'formal' mediation. Any dispute should be resolved between the promoter's agent and the contractor if possible. However, compromise should be guarded against, as an honest opinion of the merits of a situation should not be compromised without the authority of the party being represented. Where the agent has sought involvement of the principal in this way, they are using negotiation as an informal mediation technique.

Therefore, disputes which cannot be resolved in this way because they compromise the integrity of the representative(s) or because to do so would be misleading and even fraudulent, are disputes that may proceed to resolution by 'formal' mediation. Whether they should do so may depend on the type of dispute. The CDRG states that:

> 'The ... mediation service would be suitable for the less complex type of building project where disputes arise primarily in relation to quality of workmanship and materials and to variation in prices. Where there is a significant structural and or building services input, the Dispute Adviser procedure would be appropriate or, on very large projects, the Dispute Review Board.'[15]

The dispute adviser is not initially a mediator, although that role may transpire. The dispute review board is not mediation as such and should the dispute remain unresolved, it may still be mediated. The use of the dispute adviser and dispute review board is an attempt to provide a neutral overseer of the contractual issues; the traditional role of architect and engineer is often perceived, especially by contractors, as biased and not independent. The CDRG anticipate that all disputes might lead to mediation but some may have an intervening process.

Naughton suggests that nearly any dispute might be resolved by mediation:

> 'I have been impressed by how well mediation which looked hopeless actually worked. I was also impressed by the scale of some disputes which have been resolved by mediation. Not only very large cases involving millions of dollars but multiple disputes, such as the run of asbestosis and other related claims ... which may involve resolving or at least processing more than 100,000 claims.'[16]

This does not refer specifically to construction but gives a personal view of the scale and complexity with which mediation can cope. However, there is a view[17] that mediation is suitable for small disputes, and for larger disputes where legal positions are very simple, unascertainable or irrelevant. Mediation is most likely to succeed if the dispute involves primarily factual matters and or economic matters and where the parties are committed to an honest attempt to resolve the dispute[18]. It is less likely to succeed where the dispute involves a precedent. In this type of dispute, early legal advice can be invaluable.

Latham indicates that mediation is not entirely suitable for most disputes which arise on site, because he believes a speedy decision should be reached. This implies that mediation cannot offer a speedy solution, but this would be strongly contested by many advocates of mediation.

Opinions differ on which disputes are most suitable for mediation. Mediation has the potential to resolve almost any dispute, subject to the parties' commitment. However, there are some disputes which are unlikely to be the subject of mediation, such as unreasonably withholding payment of an amount certified by an independent advisor, or a clear and deliberate breach of contract, or a pure matter of law, such as the interpretation of a contract clause. Even with these mediation may not be totally excluded because although one party may have an unanswerable case, that party may wish for other reasons not to proceed to arbitration or litigation.

The utilisation and success or otherwise of mediation is far more dependent on the disposition of the parties than the nature of the dispute. Some observers[19] of ADR believe that the most successful cases are where the parties themselves suggest a solution. For this to occur settlement must be in mind, there must be a number of solutions that can be offered and the parties must not be entrenched in one view of the dispute. There must be an enduring commitment to resolving the dispute.

Circumstances where mediation may not work

Examples of where mediation may not work include disputes where:

- The dispute is centred more on law than on fact and established precedent strongly favours one party over the other. In those circumstances litigation, including summary judgment under RSC Order 14 (CPR Part 24), is more appropriate.
- One party wishes to delay the resolution of the dispute for as long as possible. If the dispute is based on a contract, the parties must be mindful of the limitation period. Under simple contracts, the limitation period operates for 6 years from the date of the contract breach occurring. If the contract is under seal the limitation period is 12 years. It would not be advantageous to commence mediation around the time when the limitation period was running out, unless the limitation period was preserved.
- Either one or other, or even both, the parties are not acting in good faith, are happy to exaggerate or even lie to a mediator and have no real commitment to resolve the dispute. Of course, neither party will recognise this in themselves and therefore it is the other party who may view one party in this light. Such a perception may arise because of the very existence of the dispute and if this were prevalent, mediation would seldom be successful.

- One party believes that litigation will be a complete vindication of their position. Parties left to their own devices may well believe this and independent advice regarding the legal position should be sought.
- There is inequality of bargaining position between the parties. This may, but does not necessarily manifest itself in their respective sizes. A small subcontractor could mediate a dispute with a large main contractor if the business affairs of each created a close dependency and if, for instance, at the time of the mediation the subcontractor could rely on an enhanced position because of his importance to the main contractor on other projects. Even where long-term partnerships are developed, differences will emerge and need resolution and the inequality of bargaining position may be secondary to the desire to maintain a working relationship.
- The position of one of the parties is strongly influenced by a particular individual in the organisation who has a position to protect, has perhaps made a mistake and is unwilling to recognise the error and can wrongly use his rank within the organisation to maintain inappropriate litigation.
- One or other of the parties has developed a culture of late or non-payment.
- One of the parties lacks the resources or the money to face their responsibilities under a particular contract but is insufficiently mature to be open about it and is happy to manipulate litigation to conceal the weaknesses.
- One party is answerable to the district auditors, in the case of a public authority, or, if an insurance-backed client, the perceptions of the professional indemnity insurers need to be satisfied. The chant frequently arises that in the absence of a properly negotiated settlement public auditors and insurers require a litigated solution. Mediation has about it the whiff of horse-trading, whereas public auditors and insurers are looking to establish rights and obligations (the simple quest for right and wrong) and are not seeking a commercial deal. In other words, the solution has to be rights based rather than interest based. What such people choose to forget, however, is that most disputes, even if litigated, do result ultimately in a deal being struck which is often lacking in finesse and owes nothing to pre-existing contractual rights and obligations. The present attitude of insurers will need to alter radically to provide further momentum to the growth of ADR in the UK, where insurers run many defences. Anecdotal evidence suggest insurers are now more positive to mediation.
- The dispute is one where the creation of precedent is desirable, for example, class actions in personal injury claims or where questions of public or administrative law are in issue on judicial review applications.
- The disclosure of confidential information, albeit on a without prejudice basis, and a failed attempt at ADR may put a party at a psychological disadvantage in subsequent litigation or arbitration proceedings, as well as raising complex questions of legal professional privilege.

The above list touches on some of the factors that may mean mediation is not the best option and indeed may not work. Nevertheless, the existence of one or more of these factors does not necessarily mean that mediation should be excluded immediately, but careful consideration should be given before embarking on a route which has no enforceable outcome without agreement.

Advantages and disadvantages of mediation

Most of the advantages put forward for mediation revolve around the following themes:

- costs (not necessarily insubstantial) are appreciably lower than with other methods of dispute resolution but not for very small disputes in County Court arbitrations or even the 'fast track' procedure
- speedier than arbitration and litigation
- less adversarial
- avoidance of legal language
- freedom from the restrictions of legal procedure
- maintains parties' hands-on involvement and hence higher level of satisfaction with outcome
- maintenance of continued business relationships
- confidentiality.

Disadvantages of mediation are:

- disclosure of parties' possible trial position (and weaknesses) should mediation fail
- equitable settlements depend on full discovery, thereby precluding early mediation and cost saving
- non-binding nature of settlement
- possible use as a delaying tactic
- quick solutions are more prone to error and can be unfair
- uncertainty as to privilege of disclosures in any subsequent legal action
- inequality of bargaining position and representation – e.g. a litigant in person and expensive insurer appointed lawyers.

The advantages and disadvantages listed above are not absolute because many factors will vary from one dispute to another and the level of advantage or disadvantage will vary appreciably. Some of these considerations are looked at here.

Cost savings

How much money might be saved by using mediation can only be conjecture. However, the earlier the mediation the greater the chance for ADR to be cost effective. One cannot re-run a dispute using a different approach, in order to compare the outcome. Also, the whole cost of the dispute resolution process (including key personnel) needs to be taken into account, not just the settlement figure which may be only a fraction of the total. The opportunity cost of a delayed settlement also needs to be assessed. Mediation will tie up key personnel for shorter periods than litigation and is less stressful, but there is no guarantee that costs will be saved; mediation may be only the forerunner to litigation or arbitration, the very processes the claimant should be attempting to avoid. Hence the costs of mediation may become an additional cost to those incurred later because an additional layer of dispute resolution has been introduced.

Speedier

Mediation frequently offers a speedy solution to a problem but is not always quicker than arbitration or litigation if 'documents only' arbitration is used or a short legal point is taken in the courts. Therefore with time, as with cost, there is no guarantee that savings will be made. Although delay, pre Woolf, has been inherent in the litigation process, the problem is sometimes exaggerated. Indeed, it may prove beneficial rather than a problem. Potential benefits include the opportunity for raising funds, and a time for reflection, negotiation and mediation.

Compared with adjudication, mediation would seem to have little, if any, time advantage. It is also possible that mediation has less time advantage over some arbitrations where the latter are subject to the 1996 Arbitration Act and are short order points such as section 39 applications. It must also be remembered that mediation may be commenced as a tactic for delaying proceedings in arbitration and/or litigation.

Less adversarial

The non-legal forum of mediation allows the participants to move away from the inherently adversarial procedures of arbitration and litigation. In arbitration and litigation each side tends to become defensive rather than look at the other side's point of view and seek solutions. The emphasis is on formalised procedures and use of legal language; these can be dispensed with when using mediation and the participants can design the procedure to suit their needs and use everyday language.

The Arbitration Act 1996 has gone a long way towards overcoming such problems in arbitration by providing a framework that enables great flexibility in determining the process. As for litigation, Lord Denning once said that it is war. War is about beating the other side and there is a winner and a loser, or more likely two losers. Fisher, Ury and Patton[20] do not entirely agree with that view as they believe the judge writes the judgment not in terms of winning and losing but explaining why and how the decision has been arrived at. However, this is probably an academic nicety for most litigants. Mediation, however, is not adversarial and is more about persuasion and negotiation. It seeks to make both parties feel that they are winners.

Parties maintain involvement

Because the mediation process is designed for the participants they are not intimidated by overly formal settings in which they have little control or where the lawyers and experts take over and run the show (usually with the parties' connivance). Mediation is conducted with the parties themselves central to the process of finding a solution (perhaps a little discomforting). This can lead to a higher level of satisfaction with the settlement than an imposed decision would bring and the parties may be able to maintain a business relationship. This will not always be the case because any dispute tends initially to weaken a business relationship, but staying in contact and continuing a dialogue do improve the chances of it.

Confidentiality and privilege

Mediation like arbitration is a private forum and consequently the detail of the dispute is not widely broadcast. This is a benefit for those with business interests to protect (such as in software disputes) but disputants do not always want confidentiality. Some foster the image of the small business being exploited by the large organisation, with publicity actively sought. Mediation and arbitration will not always remain confidential, anyway, as the parties may pass on information in order to secure publicity. Even where confidentiality clauses apply, leaks are still possible and secrecy can be difficult to achieve. Until the courts or the legislature provide a definitive answer, it cannot be said categorically that mediation proceedings are protected by privilege, although public policy would indicate so. If the proceedings are not privileged it might mean that the parties are inhibited or that some disclosure may ultimately prejudice a party's case.

Disclosure

One party may use mediation as a ploy to secure knowledge of the other party's likely stance, although cost may ordinarily militate against such cynical use of mediation in most disputes. Regardless of the motivation to use mediation, the disclosure of information is not likely to be a problem, particularly as the courts have long fostered the early exchange of experts' reports and post Woolf there will be greater reliance on pre-action disclosure.

Equitability and fairness

If speed is of the essence in resolving a dispute all documents may not be discovered and courts too are departing from blanket discovery as seen in the CPR. The mediator will not necessarily seek a solution based on the merits of the case, but a solution with which the parties can live. Fairness may be achieved without the whole case being argued. However, one can only know that it has been achieved after the whole case for both sides has been put. In other words, the settlement that is sought may better be seen as expedient, rather than fair.

Non-binding nature of mediation

Because mediation is voluntary it is non-binding unless the settlement becomes a legal document. When considering the time spent arriving at a proposal in evaluative mediation it may seem that binding adjudication or a binding expert determination is preferable.

Conciliation schemes

One of the more enterprising forms of dispute resolution developed during the 1990s is conciliation as understood by the Institute of Civil Engineers and this has been incorporated into the standard form contracts prepared by them. Its use has been sporadic and the ICE is no longer training conciliators owing to lack of interest.

Some of the conciliation procedures currently in use are identified below, with some of the key issues briefly discussed.

ICE conciliation procedure

The ICE conciliation procedure, first devised for use with the ICE Conditions of Contract for Minor Works 1988, now serves, in its 1994 revision, the ICE Conditions of Contract 6th Edition, the ICE Design and Construct Conditions of Contract and the ICE Conditions of Contract for Minor Works. The procedure is considered in detail in Chapter 12.

ICC conciliation rules

The spirit of the ICE conciliation procedure has much in common with the ICC rules of conciliation[21], which are significant in the resolution of international construction disputes. Those rules are set out in a brisk, no nonsense manner. According to the preamble: 'Settlement is a desirable solution for business disputes of an international character.'

The resolution of disputes is by a single conciliator appointed by the International Chamber of Commerce (article 1). The party requesting conciliation applies to the secretariat at the International Court of Arbitration 'setting out succinctly the purpose of the request and accompanying it with the fee required to open the file'. As with the ICE conciliation procedure, the aim is to offer expedited resolution, although under article 3 the party requested to agree to conciliate the dispute is given 15 days to inform the secretariat whether or not they agree to the conciliation option. If they do agree, the Secretary General of the International Court of Arbitration appoints a conciliator who informs the parties of his appointment and sets time limits for the presentation by each party of their case. The process is extra judicial, although article 5 allows each party to be assisted, if they wish, by their lawyer. However, the guiding principle behind the conciliation process is that the conciliator is master of his own procedures provided he operates within the principles of impartiality, equity and justice.

There are three methods by which conciliation will come to an end:

(1) The parties sign a settlement agreement which will be final and binding and will remain confidential unless its execution or application require it to be disclosed.
(2) The conciliator produces a report noting that the conciliation has failed.
(3) One of the parties notifies the conciliator at some stage during the conciliation procedure that they no longer wish to participate.

Under article 9 the parties are responsible equally for the conciliation fees, which will be fixed by the secretariat of the International Court of Arbitration. If it subsequently becomes apparent that the sum lodged with the secretariat is no longer adequate, this will be revised by the secretariat. The parties will be notified accordingly and required to pay additional equal amounts. The parties remain equally responsible for the costs of the conciliation unless they agree on some other apportionment.

Article 10 deals with the potential problem of Med-Arb. Unless the parties agree, a former conciliator cannot act in any subsequent arbitration or litigation which involves the parties. In addition, as under the ICE conciliation procedure, the parties agree not to call the conciliator as a witness in any subsequent litigation or proceedings. Article 11 provides some safeguards so that the judge or arbitrator is not prejudiced in any subsequent proceedings by what has happened during a failed attempt at conciliation. These protections are:

- any views previously expressed or suggestions made by any party to settle the dispute shall be treated as without prejudice
- proposals made to the conciliator shall not be referred to in subsequent litigation or arbitration
- no reference can be made in subsequent litigation or arbitration to the fact that one of the parties indicated that they were ready to accept a proposal for settlement put forward by the conciliator.

UNCITRAL conciliation rules

A further well used system of conciliation is provided by the UNCITRAL Conciliation Rules. These consist of twenty articles. The rules apply (article 1) to any dispute which the parties have agreed shall be resolved in accordance with the UNCITRAL conciliation rules, provided that, of course, the rules do not contradict the applicable law in a particular jurisdiction. Conciliation is commenced in accordance with the procedures in article 2. The party requiring conciliation sends a written invitation to conciliate to the other, briefly identifying the matters in dispute. If no reply is received within a period of 30 days (article 2.4) this may be treated as a rejection of the invitation to conciliate. The party which originally proposed the conciliation may then elect to inform the other party that the request to conciliate has been withdrawn.

Article 3 states that there will be a single conciliator, although a maximum of three is permitted if the parties so agree. Appointment of a conciliator or conciliators is dealt with in article 4. If there is a single conciliator, the parties attempt to agree his appointment; if there are two conciliators, each party will appoint their own conciliator; and if there are three conciliators one will be appointed by each party and both parties will endeavour to agree the appointment of the third. Article 4.2 provides for the possible appointment of the conciliator or conciliators by an appointing organisation.

Article 5 anticipates that written statements will be submitted to the conciliator with copies made available to the other party. Under Article 5.2 the conciliator may request further written statements to clarify particular issues, together with limited disclosure of relevant documents on which the parties rely to support their case.

The role of the conciliator is defined in article 7; the status of the conciliator as an independent neutral who is there to facilitate agreement between the parties is emphasised. He can be flexible in the way he approaches matters (article 7.3), while article 7.4 permits him to be interventionist, making proposals for a settlement of the dispute at any stage during the proceedings. Article 9 again emphasises the flexibility of the conciliation rules and allows the conciliator to progress by meetings and correspondence with the parties or, indeed, by telephone conversations. Information provided to the conciliator during a conciliation remains confidential unless the party giving the information waives the confidentiality (article 12).

If the parties succeed in agreeing terms of settlement, article 13 contains a procedure for reducing those terms to a written form. In this the conciliator takes the initiative. Whenever he believes that there are terms for a possible settlement he submits them to the parties for their observations and as a result may revise the terms of possible settlement. Article 13.2 anticipates that any settlement will be reduced to written form, whether with or without the assistance of the conciliator. The finalisation of the settlement agreement brings an end to the dispute (article 13.3). The settlement agreement is binding between the parties, provided it is a legal document properly drawn up. Subject to the jurisdiction of a particular country, it may be enforced in the courts. Article 15 allows for the termination of conciliation proceedings in four situations:

(1) by the signing of a settlement agreement
(2) when it becomes clear that the conciliation process has failed and it cannot be revived
(3) by a written declaration from the parties to the conciliator that the conciliation proceedings have been terminated
(4) if one of the parties confirms to the conciliator that the conciliation has been terminated.

The use of the conciliation rules acts as a suspension of the right to bring arbitration or litigation proceedings during the period of the conciliation (article 16).

As under other conciliation schemes, costs, as defined in article 17.1, are to be a joint and several responsibility of the parties (article 17.2) unless this is modified by the settlement agreement. Article 18 makes provision for deposits to be taken by the conciliator as security against his own future costs. Unlike other conciliation schemes, which allow the conciliator by agreement to be used as an arbitrator in subsequent proceedings, article 19 of the conciliation rules states that the conciliator will have no role in any subsequent proceedings, either as arbitrator, legal representative or witness. Further, in similar terms to those used in the ICC rules of conciliation, article 20 makes it clear that any proposals, information, etc. which come to light during the course of a conciliation are inadmissible in any subsequent litigation or arbitration.

Chapter 5
Adjudication and Expert Determination

Some see adjudication/expert determination as having much to offer, whatever happens to court procedures after 26 April 1999, when streamlined by the Woolf reforms, and with the jury still out on the future of arbitration as the 1996 Act descends into legal cases and JCT Amendment 18 and JCT 98 having provided the litigation option for the resolution of disputes. Disputes often involve highly technical issues rather than legal ones and this raises the spectre of people highly trained in the law having to address and resolve technical matters. The following extract[1] epitomises the problem as seen by industry professionals:

'Sadly the last resort – action through the courts – has once again been necessary to resolve a purely technical argument. This time it was to establish why the Ness railway viaduct collapsed in 1989 and whether the downstream damage was a direct result of the failure. A Scottish court has ruled that owner British Rail was to blame on both counts but only after the judge Lord Clyde had sat through a month of complex, contradictory and sometimes ambiguous debate on scour proffered by civil engineer expert witnesses called both by BR and Inverness Harbour Trust, the company claiming the cost of the damage.

Lord Clyde's analysis of how scour attacked and felled the crossing was convincing even though he had to totally reject the evidence of competent engineers on the losing side and add an escape clause that the ruling was based on probability not scientific certainty.

But as the harbour engineer said afterwards, there must be a better way. Some form of technical arbitration should be possible as an alternative to the courtroom battle.'

Perhaps adjudication and/or expert determination can provide the answer and ensure a party is not kept out of money that may be properly due for any longer than is necessary.

Adjudication

As every specialist contractor is aware, cashflow is the lifeblood of the industry. For this reason, and well before the Housing Grants, Construction and Regeneration Act 1996 became law, adjudication was introduced into standard form subcontracts such as DOM/1 (clause 24) (discussed in

Chapter 10) and DOM/2. Rather than have to pursue matters through the courts, subcontractors would be able to obtain prompt and cheap justice with a quick review of main contractor set-offs by an experienced neutral person. Sir Michael Latham, in his review[2] of the construction industry, came out strongly in favour of adjudication even if he did not define it. Latham wished to see the role of adjudication extended beyond the narrow one largely found at the time of the review. Already, in its core clauses, 90–93, the New Engineering Contract 2nd Edition (strongly endorsed by Latham) had widened the range of circumstances in which adjudication might be used. Latham made five recommendations in paragraph 9.14 of his review:

'1. There should be no restrictions on the issues capable of being referred to the adjudicator, conciliator or mediator, either in the main contract or subcontract documentation.
2. The award of the adjudicator should be implemented immediately. The use of stakeholders should only be permitted if both parties agree or if the adjudicator so directs.
3. Any appeals to arbitration or the court should be after practical completion, and should not be permitted to delay the implementation of the award, unless an immediate and exceptional issue arises for the courts or as in the circumstances described in (4).
4. Resort to the court should be immediately available if a party refuses to implement the award of an adjudicator. In such circumstances, the courts may wish to support the system of adjudication by agreeing to expedite procedures for interim payment.
5. A training procedure should be devised for adjudicators. A Code of Practice should also be drawn up under the auspices of the proposed Implementation Forum.'

The main tenor of these recommendations was reflected in the Department of the Environment's consultation paper, *Fair Construction Contracts*[3], and subsequently became a part of the Housing Grants Construction and Regeneration Act 1996. In a Department of the Environment news release[4] prior to the Act, the political will in favour of adjudication was reflected in the following:

'Adjudication
There will be a statutory entitlement for any party to a court action to seek independent adjudication. In the event of a dispute, adjudication will be a right rather than a requirement. The terms and scope of this will follow closely the provisions of the Arbitration Bill, except that certain provisions will be modified or supplemented to provide for "fast track" dispute resolution to meet the specific needs of the construction industry.'

This extract illustrates the desire for mandatory provisions but with no compulsion to use adjudication should a dispute arise.
According to the Construction Industry Council[5]:

'Adjudication is a *quick fix* . . . It is a means of *rough justice* to enable a quick decision, which prevents the escalation of disputes and the holding up of work. It is a valuable process so long as its *quick fix* nature is recognised and respected.

Unfortunately, the Government seems unwilling or unable to understand adjudication for what it is. In the draft proposals for the Scheme, adjudication is treated almost as if it were arbitration.'

The Housing Grants, Construction and Regeneration Act 1996 (HGCRA) received the royal assent in July 1996, although it did not become active law until 1 May 1998. Among the provisions, some of which deal with the right of payment, section 108 imposes mandatory adjudication provisions on a wide range of construction contracts, defined in convoluted terms in sections 104 and 105. In the absence of a complying procedure, the parties will be bound by the Scheme for Construction Contracts Regulations, brought into law by means of a statutory instrument. Adjudication under this Act is discussed in Chapter 9.

The practical application of particular adjudication clauses has had its problems. First, many main contractors in the past have deleted all reference to adjudication from subcontract documents. Second, there have been a number of cases dealing with adjudication and many of these have been unhelpful to specialist contractors.

Apart from the legal problems adjudication poses, and the need to interpret and implement particular expert determination or adjudication provisions, Latham failed to address the practical problems satisfactorily in his report *Constructing the Team*[6]. These include:

- Should adjudication be the exclusive remedy in construction contracts prior to practical or substantial completion of building works?
- Should adjudication be final and binding on the parties until practical or substantial completion and only reviewable through the courts or arbitration afterwards?
- Who pays for the adjudicator? Is the adjudicator a project expense or only to be reimbursed when actually required to act?
- How is the adjudicator to be kept informed? Should the adjudicator (if appointed at contract commencement) receive copies of all project documentation including 'contractual' correspondence or simply periodic progress minutes?
- Should the adjudicator read the information provided to him when supplied or only if a dispute arises?
- What are the adjudicator's legal liabilities?
- What if the adjudicator is reluctant to make a rapid decision on complex questions which might subsequently prove to be incorrect?

Of course, the HGCRA has, in part, provided its own answers to some of the above, whether expressly or by omission.

Clients criticise litigation and arbitration, the traditional lawyer-led methods of dispute resolution. They are perceived as too long and too expensive. Lord Woolf may change all that, but are lawyers, clients and all those who benefited from the 'old' system prepared to let reform work? The quotation from *New Civil Engineer* at the beginning of this chapter demonstrates how unconvinced the construction industry is that lawyers (and rightly so) are the appropriate people for the resolution of what are usually technical disputes. Many situations on site require technical assessment; interpretation of drawings and specifications or review of workmanship, rather than the application of the law, are crucial. Rather than simply allow these problems to multiply by practical or substantial completion stage, when they may cease to be minor sores and become festering wounds, many construction professionals consider better project management requires an experienced neutral professional to be on hand to provide solutions to problems as and when they arise on site, with the possibility of a review of the neutral's decisions (if one of the parties so wishes) once major construction work has been completed. This was the underlying message of Sir Michael Latham's reports[6].

Instant justice may be more appropriate on construction projects. For instance, in the construction of a new dam, certain disputes may need to be resolved early. Some parts of the works may be submerged below many metres of water and not capable of reassessment at practical or substantial completion.

Even prior to the HGRCA 1996, many bespoke contracts and certain standard form contracts included a provision for expert determination or adjudication, which the prudent litigation lawyer always had to treat as a potential impediment to the commencement of High Court proceedings. Adjudication was used on the Dartford Crossing, while on other complex projects such as the Channel Tunnel, the Second Severn Crossing and the Department of Transport's A35 trunk road scheme, the collective wisdom of a three or more member dispute review board (discussed in Chapter 6) was available. Outside the UK, disputes arising on the Hong Kong Airport Core Programme projects (discussed in Chapter 3) could go through four stages: a decision of the engineer, mediation, adjudication and finally arbitration.

Adjudication is a system whereby the parties to a contract agree, usually on entering into the contract, that all disputes arising under or possibly out of the contract (the HGCRA 1996 is restrictive on this point so as to exclude collateral claims for negligence, misrepresentation etc.), or certain prescribed areas of dispute, shall in the first instance be referred for the binding decision of an adjudicator, which will remain effective until replaced by litigation, arbitration or agreement. The adjudicator is either named in the contract, agreed by the parties or, in the absence of agreement, nominated by an appointing body. Adjudication is flexible; the decision of the adjudicator may be final and binding on an interim basis (except where it can be attacked on the grounds of fraud, procedural unfairness, manifest incompetence or lack of jurisdiction) in regard to some, or even all, issues. Expert determination is usually final and binding on all issues, particularly valuation.

Expert determination is a well known principle in commercial property documentation, oil/gas agreements and share sale agreements. In some senses it is close to arbitration, but in its simplest form it exists where parties appoint a valuer to determine a fair price for something. It operates where an expert opinion is required rather than an award based on submissions from the parties. The expert's decision cannot be challenged in the same way as an arbitral award through the courts, although it may sometimes be set aside because of fraud, collusion, breach of natural justice or want of jurisdiction.

But to return to adjudication a big problem is posed by initial challenges to the adjudicator's jurisdiction. What if one party disputes the existence of a contract, or claims are brought in negligence, misrepresentation, recission or restitution, or following termination of the contract by breach? The HGCRA presupposes a contract to exist. What if the adjudicator suffers threats, usually from bullish lawyers, to his jurisdiction? With the exception of the Official Referees and Solicitors Association (ORSA) rules, now the TecSA rules, which allow an adjudicator to rule on his jurisdiction, the other institutional rules are silent. Sections 30–31 of the Arbitration Act 1996 address the point and so provide guidance by analogy. Following the decision in *Harbour Assurance* v. *Kansa General International Insurance* (1993), there is a persuasive argument the adjudicator probably has jurisdiction to investigate and decide if he is empowered to act.

The possible value of adjudication has been recognised for a number of years but on the basis legalism is not allowed to take over:

> 'Its value is in making both sides think again before persisting with an arbitrary decision which can only generate a dispute. I have made recommendations for adjudicators when asked to do so, but afterwards people appointed phone me to complain that they haven't had anything to do! My experience is that the mere knowledge of there being an adjudicator in the background to whom disputes can be referred makes the whole operation run much more smoothly.'[7]

On the face of it and from a practical point of view, adjudication has many virtues which are not shared by litigation or arbitration. Adjudication avoids the enforced legalism of the *White Book*, the *Green Book* (and possibly the new Civil Procedure Rules) the Arbitration Acts 1950–1979, and to a lesser extent the Arbitration Act 1996. In broad terms, provided that the adjudicator is not fraudulent, unfair, manifestly incompetent or acting outside his brief, he is master of his own procedures, subject to any veto vested in the parties. Within the boundaries set to his appointment, he is able to provide a system of summary justice. Some commentators say that an adjudicator need not respect the principles of natural justice but this is perhaps overstated.

The courts are reluctant to lay down hard and fast rules as to what constitutes natural justice. In *Wiseman* v. *Borneman* (1971), Lord Reid stated[8]:

> 'Natural justice requires that the procedure before any tribunal which is acting judicially shall be fair in all the circumstances, and I would be sorry

to see this fundamental principle degenerate into a series of hard-and-fast rules.'

Many of the cases on breaches of natural justice relate to the way in which public bodies have exercised their powers. In *R.* v. *Lord President of the Privy Seal ex parte Page* (1992)[3] the House of Lords held that a lecturer, made redundant by a university, was not entitled to challenge the findings of the internal university enquiry in court merely for error in fact or law. The university's decision could only be reviewed if it was in excess of their powers or an abuse of power. Just to confuse the picture, in *R.* v. *Disciplinary Committee of the Jockey Club ex parte Aga Khan* (1993) the Court of Appeal declined to intervene in proceedings at the Jockey Club on the basis that only private and not public rights were involved.

However, the responsibilities of an adjudicator may be less onerous than those of a judge or arbitrator. Under the HGCRA, although the adjudicator is supposed to be impartial (section 108(2)(e)), the procedure is not laid down, apart from the guidelines in subsection (2). It is not expressly stated that an adjudicator has to observe the rules of natural justice. Although the principles of natural justice may be less onerous than in the case of an arbitrator or judge, it is worth bearing in mind the words of Tucker LJ in *Russell* v. *Duke of Norfolk* (1949)[9]:

'There are, in my view, no words which are of universal application to every kind of enquiry and every kind of domestic tribunal. The requirements of natural justice must depend on the circumstances of the case, the nature of the inquiry, the rules under which the tribunal is acting, the subject matter that is being dealt with, and so forth. Accordingly, I do not derive much assistance from the definitions of natural justice which have been from time to time used, but, whatever standard is adopted, one essential is that the person concerned should have a reasonable opportunity of presenting his case.'

A number of cases have considered the status of adjudication. At present, lawyers cannot fully define the jurisprudential nature of adjudication. What case law exists is relatively small, leaving every possibility for a disgruntled party to an adjudication to seek recourse to the courts in appropriate circumstances. The case of *Cape Durasteel Ltd* v. *Rosser and Russell Building Services Ltd* (1995)[12] indicates that the courts will look at the true import of a dispute clause rather than automatically characterise it as an adjudication clause. In *Cape Durasteel*, the Official Referee decided that what purported to be an adjudication clause in a management contract derivative was in fact an arbitration clause, given that the result was final and binding. In *Cape Durasteel*, counsel for the defendant referred to a 'checklist'[10] of the attributes of arbitration which might prevent the imputation that a particular clause was an adjudication clause:

'The agreement pursuant to which the process is, or is to be, carried on (the procedural agreement) must contemplate that the tribunal which carries on the process will make a decision which is binding on the parties to the procedural agreement.

The procedural agreement must contemplate that the process will be carried on between those persons whose substantive rights are determined by the tribunal.

The jurisdiction of the tribunal to carry on the process and to decide the rights of the parties must derive from the consent of the parties, or from an order of the court or from a statute the terms of which make it clear that the process is to be an arbitration.

The tribunal must be chosen, either by the parties, or by a method to which they have consented.

The procedural agreement must contemplate that the tribunal will determine the rights of the parties in an impartial manner, with the tribunal owing an equal obligation of fairness towards both sides.

The agreement of the parties to refer their disputes to the decision of the tribunal must be intended to be enforceable in law.

The procedural agreement must contemplate a process whereby the tribunal will make a decision upon a dispute which is already formulated at the time when the tribunal is appointed.'

Some commentators suggest that an adjudicator, although specifically described as not acting as an arbitrator, may be subject to the Arbitration Act whenever he makes a *legal* decision. Cases concerning expert determination provide some guidance. In the past, many contract clauses which call for the appointment of an expert state that the expert shall sit as such, not as an arbitrator and not subject to the Arbitration Act. However, there can be a fine dividing line between an adjudicative and an arbitral role. In the words of Ronald Bernstein QC[11]:

'A contract may provide that disputes arising under it are to be resolved by some third person acting not as arbitrator but as an expert ... The procedure involved is not arbitration and the Arbitration Acts do not apply to it.'

Unfortunately, the application of such a broad principle may not be as simple as that. According to Mustill and Boyd[12]:

'The way in which the reference is described in the agreement to refer is not conclusive as to the character of the proceedings. Thus, even an explicit agreement that a matter be dealt with by arbitration does not mean that the parties intended the proceedings to be the type of arbitration which is the subject of the Arbitration Acts or the common law of arbitration. For example, the use of this word is consistent with an intention to invoke a process which involves a decision by an impartial body, but not one which is binding in law.'

Mustill and Boyd further comment[13]:

'...provisions purporting to exclude the Arbitration Acts, or providing that the tribunal shall "sit as experts and not as Arbitrators" could not (in the absence of very strong indication to the contrary) be regarded as consistent with an intention to refer disputes to arbitration.'

In the case of statutory adjudication, the characterisation is provided by HGCRA itself.

In one helpful overseas case, *Sports Maska Inc.* v. *Zittrer* (1988), which pre-dated *Cape Durasteel*, the Supreme Court of Canada considered the language used was not decisive. What the parties describe as an expert determination may be, on objective analysis, arbitration. Such case law nuances provide great potential difficulties for formal adjudication procedures brought into construction contracts by legislation.

The adjudicator

Ordinarily an adjudicator will have a formal notice of appointment with the parties (as under the adjudicator's contract for use with the New Engineering Contract), whether they be client and main contractor or main contractor and subcontractor or client and consultant. The appointment will include the adjudicator's responsibilities and fix the method for his remuneration. A very detailed procedure may be set out for the conduct of an adjudication (guidance on whether written statements of case are to be submitted, inspection of work to be made by the adjudicator, oral examination of the parties by the adjudicator, etc.) or the appointment may be set out in very general terms.

Unless the adjudicator's appointment contains a specific disclaimer of responsibility, although the grounds for setting aside his decision are limited, the adjudicator has a duty to the parties to use reasonable skill, care and diligence. Unlike a judge or arbitrator, he does not enjoy immunity from claims. His position is nearer to that of the traditional certifying engineer or architect under construction contracts, albeit with a major distinction. He is retained by both parties and owes a clear duty to both. A number of legal decisions during the course of the nineteenth and early twentieth centuries identified the status of valuers and certifiers under standard form and other building contracts. These assist an analysis without providing a full explanation of the adjudicator's legal role. The nineteenth century cases suggest that the certifier under standard form building contracts enjoyed a sort of judicial immunity of the type usually associated with arbitrators and judges. For instance, in *Hickman & Co* v. *Roberts and Others* (1913) the architect delayed payment to the contractor on the specific instructions of the employer. In holding that it was improper for the employer to interfere with the certifier's discretion under the contract, the House of Lords made a number of interesting observations. In the words of Lord Alverstone[14]:

'It is therefore very important that it should be understood that when a builder or contractor puts himself in the hands of an engineer or architect as arbitrator there is a very high duty on the part of that architect or that engineer to maintain his judicial position.'

In an earlier case, *Chambers* v. *Goldthorpe* (1901), a building owner unsuccessfully sued the construction professional for negligent certification, which resulted in overpayment to the contractor. The court held that, when certifying, the constructional professional had to act impartially as between employer and contractor. Although not an arbitrator in the strict sense of the term, an architect was, in the words of A.L. Smith MR[15]:

'In the position of a person who had to exercise functions of a judicial character as between two parties, and therefore was not liable to any action for negligence in respect of what he did in the exercise of those functions.'

This remained the position until two key decisions of the House of Lords in the 1970s. In the first, *Sutcliffe* v. *Thackrah* (1974), Sutcliffe had employed Thackrah, an architect, to design a house. Subsequently Thackrah was appointed as supervising architect for the works, during the course of which he issued interim certificates to the contractor. After the original contractor's employment had been determined, the contractor went into liquidation. Sutcliffe brought an action against Thackrah for damages for negligent certification and/or supervision of the works. The Official Referee held that Thackrah had negligently over-certified sums due to the contractor. The Court of Appeal reversed this judgment; Thackrah was acting in an arbitral capacity and therefore enjoyed the benefits of *judicial* immunity. Sutcliffe appealed to the House of Lords. In reversing the Court of Appeal's decision, the House of Lords held that in issuing interim certificates an architect did not, in the absence of special agreement to the contrary, act as an arbitrator between the parties. He was therefore under a duty to act fairly in making valuations and in the absence of competence, such an architect was clearly liable in the tort of negligence. In the words of Lord Salmon[16]:

'It is well settled that judges, barristers, solicitors, jurors and witnesses enjoy an absolute immunity from any form of a civil action being brought against them in respect of anything they say or do in court during the course of a trial. This is not because the law regards any of these with special tenderness but because the law recognises that, on balance of convenience, public policy demands that they shall all have such an immunity.

Since arbitrators are in much the same position as judges, in that they carry out more or less the same functions, the law has for generations recognised that public policy requires that they too shall be accorded the immunity to which I have referred. The question is – does this immunity extend beyond arbitrators properly so called, and if so, what are its limits?'

The answer was that it did not.

In a more recent case, *Arenson* v. *Casson Beckman Rutley & Co.* (1977), Archy Arenson was the controlling shareholder and chairman of a private company. His nephew, Ivor Arenson, subsequently went into business with Archie and was given a number of shares. These were held, inter alia, on the basis that on termination of Ivor's employment, the latter would sell his shares to Archie at a fair value, as determined by the auditors of the company. The key phrase was 'whose valuation acting as experts and not as arbitrators shall be final and binding on all parties'. The auditors subsequently valued the shares at £4,916 13s 4d, whereas it later transpired that the shares were actually worth £29,500. The nephew brought proceedings against the auditors claiming damages for negligence. The trial judge and the Court of Appeal held that since the auditors had been performing an arbitral or quasi-judicial function they were immune from any liability for negligence. The nephew appealed to the House of Lords. The House of Lords held that a person or persons whose appointment was expressly stated to be *as expert and not as arbitrator* rendered him or them liable in the tort of negligence for any incompetence in the carrying out of their duties. In the words of Lord Simon[17]:

'There may well be other indicia that a valuer is acting in a judicial role, such as the reception of rival contentions or of evidence, or the giving of a reasoned judgment. But in my view the essential pre-requisite for him to claim immunity as an arbitrator is that, by the time the matter is submitted to him for decision, there should be a formulated dispute between at least two parties which his decision is required to resolve. It is not enough that parties who may be affected by the decision have opposed interests – still less that the decision is on a matter which is not agreed between them.'

Lord Wheatley identified the following features as characteristic of an arbitral appointment:

'(a)   there is a dispute or difference between the parties which has been formulated in some way or another;

(b)   the dispute or difference has been remitted by the parties to the person to resolve in such a manner that he is called upon to exercise a judicial function;

(c)   where appropriate, the parties must have been provided with an opportunity to present evidence and/or submissions in support of their respective claims in the dispute; and

(d)   the parties have agreed to accept his decision.'[18]

Lord Salmon summed up the position as follows[19]:

'an expert may be formally appointed as an arbitrator under the Arbitration Acts, notwithstanding that he is required neither to hear nor read any submissions by the parties or any evidence and, in fact, has to rely on nothing but his examination of the goods and his own expertise. He, like the valuer in the present case, has a purely investigatory role; he is performing no function even remotely resembling the judicial function save that he finally decides a dispute or difference, which has arisen between the parties . . .

I find it difficult to discern any sensible reason, on grounds of public policy or otherwise, why such an arbitrator with such a limited role, although formally appointed, should enjoy a judicial immunity which so-called quasi-arbitrators in the position of the respondents certainly do not . . .'

Adjudication and expert determination clauses

From the discussion above it can be seen that adjudication and expert determination clauses need to be examined carefully to decide their true meaning and effect. It may be that a particular clause:

- excludes the jurisdiction of the courts, or of an arbitrator, on an interim basis with the adjudicator's decision subsequently open to review; or
- provides a full and final binding decision on the parties in which event the clause, however described, is either arbitration or expert determination; or
- although described as an adjudication provision is in fact an arbitration clause to which the Arbitration Act applies.

The English courts have considered expert determination and adjudication procedures in a number of cases as have courts in the USA and Australia. A notable example of a graduated disputes procedure was the subject of court proceedings in *Channel Tunnel Group Limited* v. *Balfour Beatty Construction Limited* (1993), a decision of the House of Lords. The contract was based on ICE/FIDIC with a number of significant changes. Under Clause 67 (disputes) there was provision for reference to a panel of three experts whose decision had to be unanimous, with ICC arbitration in Brussels to follow, if necessary. A dispute arose over the pricing of a variation for the cooling system. In the absence of agreement with the contractor, the employer fixed the rate to apply. The contractor found this unacceptable and threatened to suspend work on the cooling system. The employer issued proceedings for an injunction, the matter at that stage not having been referred to the panel of experts. For whatever reason, the role of the engineer under the contract had been substantially undertaken by the employer.

In November 1991, Evans J sitting in the Commercial Court decided that he would have granted an injunction and refused a stay of proceedings to arbitration under section 1 of the Arbitration Act 1975. In January 1992, the Court of Appeal decided there should be a stay of action under section 1 of the 1975 Act but refused an injunction on the grounds that there was no power to grant one under section 12(6)(h) of the Arbitration Act 1950. In

January 1993, the House of Lords decided there should be a stay under the inherent power of the court, so that it was unnecessary to decide whether section 1 of the 1975 Act applied. The House of Lords agreed that there was no power to grant an injunction under section 12(6) of the Arbitration Act 1950, although there was a general discretion under section 37(1) of the Supreme Court Act 1981. Such power was not to be exercised in the instant circumstances.

As previously mentioned, development agreements often contain provision for expert determination on questions of valuation where particular disputes will be settled by an expert expressly described as not fulfilling the role of an arbitrator or being subject to the Arbitration Act. In one Court of Appeal decision, *Norwich Union Life Insurance Society* v. *P & O Property Holdings Limited and Others* (1993), the court held that where a dispute had been submitted to an expert for determination, a court should not make a ruling on any issue of law arising in the course of the determination except by the consent of the parties. Previous decisions to the contrary were over-ruled. In most cases it will not be possible for the courts to intervene, to reverse the conclusions reached in an expert determination. Australian courts appear to have adopted a similar stance. Cole J held in *Triarno Pty Limited* v. *Triden Contractors Limited* (1992) that the courts had little scope to intervene in an expert determination, the effect of which was final and binding. However, of course it will still be open to the parties (subject to a specific disclaimer in the appointment) to sue the expert for negligence on the principles developed in and after *Sutcliffe* v. *Thackrah* (1974).

In England, expert determination clauses have also been considered in *Amoco (UK) Exploration Co. and Others* v. *Amerada Hess Limited and Others* (1994) and *Neste Production Limited* v. *Shell UK Limited and Others* (1994). These decisions demonstrated just how carefully a particular clause needed to be analysed to define the extent to which the courts' powers to intervene were suspended or removed.

In *Jones and Others* v. *Sherwood Computer Services plc* (1992), the defendant company agreed to purchase the plaintiffs' shares in a particular company, on terms which included the issue to the plaintiffs of a number of new shares in the defendant company. The value of these shares was to be calculated by reference to the amount of sales as determined by a prescribed formula. The statement of the amount of sales was to be reviewed by firms of accountants representing the parties. If those firms were unable jointly to approve the statement or agree on adjustments, the matter was to be referred to independent accountants to determine and report the amount of sales, with the accountants acting 'as experts and not as arbitrators'. Further, it was agreed that their determination was 'conclusive and final and binding for all purposes'.

The Court of Appeal held that where parties had agreed to be bound by the report of an expert, the report, whether or not it contained reasons for the conclusions, could not be challenged in the courts on the grounds that mistakes had been made in its preparation unless it could be shown that the expert had departed from the instructions given to him in a material respect or that the independent expert had demonstrated bad faith.

In *Norwich Union Life Insurance Society* v. *P & O Property Holdings Limited and Others* (1993), the plaintiff had provided development funding for a shopping centre. The funding agreement required particular matters to be referred to an expert for determination. A dispute arose over whether 'completed' under the funding agreement and 'practical completion' under the building contract meant the same. Norwich Union contended that a higher standard was required under the funding agreement than under the building contract. In accordance with the funding agreement, the fourth defendant was nominated as an expert to determine this question, among other issues. The plaintiff, Norwich Union, attempted to obtain an inter-locutory injunction to restrain the fourth defendant (the expert) from pro-ceeding, pending determination by the court of all issues. The Court of Appeal held that the courts did not have a general jurisdiction to make rulings on any issue of law arising in the course of an expert determination except by consent of the parties. At first instance, counsel had relied on a dictum of Judge Paul Baker QC, sitting as a deputy judge of the High Court, in *Ponsarn Investments Limited* v. *Kansallis-Osake-Pankki* (1992). The Vice Chancellor, Sir Donald Nicholls, referred to this in his judgment[20], albeit he rejected the deputy's proposition:

'The learned judge [Judge Baker] made an observation on which [counsel] placed some reliance. In that case the judge was concerned to address, and in his judgment he rejected, a submission that, if an expert went wrong in law, he exceeded his jurisdiction and his determination was a nullity... The judge gave several reasons for rejecting this submission. He gave, as one of his reasons:
   "Second, if it is thought that some question of law requires resolution, either party can seek a declaration from the court".'

In the Court of Appeal, Dillon LJ said[21]:

'We were referred also to the decision of Hoffmann J in the case of *Royal Trust International Limited* v. *Nord Banken* decided on October 13th 1989 but unreported.'

At page 6 of the transcript he said:

'I do not think that it is right that the court has no jurisdiction to make declarations in advance of an expert's determination except with the consent of the parties.'

He considered that the court had a discretion whether or not to grant such declarations and to stay the proceedings, if necessary, pending the making of such declarations:

'But with all respect, I do not agree. The function of the expert is to make the decision and that is not the function of the court where the decision has been entrusted to the expert. It is otherwise if both parties agree – as they often do – to get a ruling from the court to determine the basis on which an expert is to proceed, and if it is practical to assist the court will do so. But here there is no such agreement.'

The question of the expert's role and responsibilities was also the subject of an interesting Hong Kong decision, *Mayers* v. *Dlugash* (1994). The plaintiff was the sole beneficial owner of shares in Far East Diversity Investments Limited (FEDI) and the defendant beneficial owner in shares in Common Seal Limited (CSL). FEDI and CSL were the registered owners of shares in Imcor Limited (I) in equal proportions. In 1992, the plaintiff and the defendant agreed in principle that the business activities, assets and liabilities of I should be distributed between them. By a deed of submission dated 30 March 1992, the parties agreed to appoint an independent third party to resolve any differences and to determine the manner in which such distribution was to be made. The deed was inconclusive as to whether or not the third party was appointed as an arbitrator or expert. Subsequently, the dissatisfied plaintiff sought removal of the arbitrator for misconduct under the relevant section of the Hong Kong Arbitration Ordinance. Both parties agreed that the judge, Kaplan J, should decide whether the third party expert was an arbitrator or expert as a preliminary issue. According to Kaplan J:

- An expert makes a final and binding decision
- The decision can only be challenged in the most exceptional circumstances such as where the expert answers the wrong question
- The expert can be sued for negligence in the absence of an agreed immunity
- The expert's determination cannot be enforced as an arbitral award.

The ways in which the courts could intervene in an expert determination were considered by Judge Bowsher QC, one of the Official Referees, in *Dixons Group plc* v. *Jan Andrew Murray-Oboynski* (1998). The defendants sold a group of companies to the plaintiff. Part of the sale price was to be an amount equal to the value of the group's net assets. This was to be determined from the group's accounts at the date of completion. If the accounts could not be agreed, the sale agreement provided that 'any disputed items . . . should be referred for final settlement to a firm of chartered accountants . . . as experts and not as arbitrators . . . and their decision shall be final and binding on the parties save in the case of manifest error'. An accountant was appointed and given a wide discretion. He heard submissions and made his valuation without reasons. The defendants challenged the decision in court, contending that it was not final and binding, on the grounds that the expert had departed from his instructions and there were manifest errors in his determination.

The judge accepted that, following various Court of Appeal decisions, the expert's decision was final in most circumstances, leaving the courts with little opportunity to intervene. In the absence of fraud or collusion, it was usually necessary to show that the expert had departed from his instruction in a material respect; that the answer was simply wrong did not suffice in most circumstances. Judge Bowsher referred to *Nikko Hotels (UK) Ltd* v. *MEPC plc* [1991][22]:

> 'If he has answered the right question in the wrong way his decision will nevertheless be binding. If he has answered the wrong question, his decision will be a nullity.'

In *Dixons*, the plaintiff had difficulties in demonstrating that the expert had prepared his valuation on the wrong basis. He could identify neither the instructions said to have been breached nor the breach itself. The proper basis for the valuation had to be derived from the terms of the agreement and the joint instructions to the expert. Neither the agreement nor the letter of instruction specified the basis of valuation. The interpretation of the agreement was left to the expert and the basis of valuation was a matter for him. The final straw was the absence of reasons. Therefore, it was difficult to demonstrate that he had been guilty of manifest error.

The construction law cases which have considered adjudication are, by and large, of limited value although there are now two reported cases on the enforcement of an adjudicator's decision referred to in the Preface. They have largely sought to interpret particular clauses in standard form sub-contracts calling for the adjudication of set-offs. Perhaps the most instructive is *R.M.C Panel Products Limited* v. *Amec Building Limited* (1993). The main contractor failed to comply with the adjudicator's decision and the matter came before the judge on a summary judgment/mandatory injunction application by the subcontractor. The adjudicator had been asked to consider the validity of main contractor set-offs under two contracts in the form DOM/1 and DOM/2 relating to the same overall works. The adjudicator had given reasons for his decision and the judge, on reading these, was far from convinced as to their validity and felt that the decision might well be overturned in a subsequent arbitration. He therefore side-stepped the subcontractor's application by holding that the adjudicator had exceeded his jurisdiction by giving reasons, which was in excess of his powers under clause 24.3.1 of DOM/1.

There is reason to believe, because of the way in which the courts have approached expert determination provisions, that the English courts would similarly 'lock out' court proceedings if a contract contained a properly constituted adjudication provision. First, the courts have an inherent power to stay any action which they consider should not be allowed to continue. Second, certain commentators have identified statements in Dunn LJ in *Northern Regional Health Authority* v. *Derek Crouch Construction Company Limited* (1984) as supportive[23]:

> 'Where parties have agreed on machinery ... for the resolution of disputes, it is not for the court to intervene and replace its own process for the contractual machinery agreed by the parties.'

A similar position was adopted by Kerr LJ in *Tubeworkers Limited* v. *Tool Construction Limited* (1985)[24].

Expert determination or adjudication may pose enforcement problems although the cases referred to in the Preface are bringing some clarity in the context of the HGCRA. If the unsuccessful party ignores or refuses to abide by the determination, there are two obvious routes by which to attempt enforcement. First, it may be possible to obtain a court injunction to uphold the expert's determination or adjudicator's finding. Second, and particularly in the context of construction industry disputes, where the adjudication usually relates to monies which have been set-off and which the adjudicator decides should be paid in favour of the subcontractor, can the adjudicator's award be used as evidence of entitlement to summary judgment or an interim payment.

Many of the decisions which relate to the very specific wording in standard form subcontracts are complex in their legal analysis. Even if the contractor's right to set-off monies may be subject to adjudication, most construction contracts traditionally preserved the contractor's concurrent rights in abatement without subjecting such rights to the adjudication procedures. Nowadays the requirement of prior notification of claims embraces all cross claims, including ones in abatement, if there is to be compliance with the HGCRA (section 111). In *Acsim (Southern) Limited* v. *Danish Contracting and Development Co. Limited* (1989), it was held that the contractor was entitled to defend the subcontractor's claim on the basis that the subcontractor was, because of his defective performance, entitled to a lesser sum than the one claimed. Here the subcontract used was the NFBTE/FASS/CASEC domestic subcontract (1978 revision) (the 'Blue Form').

Clause 15 required proper notice of set-off to be given by the contractor before the due date for payment. The subcontractor believed the contractor had failed to apply properly the set-off procedures in clause 15, thereby entitling the subcontractor to the release of £221,018.03, which had been withheld. The contractor contended that, regardless of any procedural breaches in making the set-off, the true value of outstanding monies owed to the subcontractor was £36,952.00. The Court of Appeal held that Clause 15 only defined the parties' rights in set-off. Under clause 13 the subcontractor was only entitled in his monthly interim payments to 'the total value of the sub-contract works properly executed' (subject to retention and any discount) within 14 days of the application. The contractor had, therefore, a right to defend a claim for an interim payment by demonstrating that the sum claimed included sums to which the subcontractor was not entitled, either because certain work had not been done or because the work was rendered worthless because of the subcontractor's breaches of contract.

The decision in *Acsim (Southern) Limited* is clearly also relevant to other subcontract forms, including NAM/SC and the 'old' DOM/1. A contractor's reliance on abatement can render futile a subcontractor's resort to adjudication of an allegedly unfair set-off. In *A. Cameron Limited* v. *John Mowlem and Company plc* (1990) the adjudicator appointed under a DOM/1 subcontract found the contractor's set-off to have been invalid. Following the contractor's refusal to abide by the adjudicator's decision, the Court of Appeal decided that the contractor could dispute the amount due to the subcontractor on the grounds that the subcontractor's valuation had included sums for work not 'properly executed'. Clause 21.4.1.1 of DOM/1 entitled the subcontractor to be paid 'the total value of the sub-contract work on site properly executed'. The doctrine of abatement served the contractor. The HGCRA has rendered this old line of reasoning redundant on modern contracts.

The decision in *Cameron* must be viewed alongside the decision of the Official Referee in *Drake and Scull Engineering Limited* v. *McLaughlin and Harvey plc* (1992), again concerning the DOM/1 form of subcontract. The subcontractor referred a disputed set-off to adjudication. The adjudicator ordered the contractor to pay £149,451 to a trustee stakeholder pending arbitration. This the contractor failed to do. During the course of the subsequent arbitration, the subcontractor sought a mandatory injunction from the courts to enforce the adjudicator's decision. One of the arguments the contractor put forward to resist the injunction application was that there was no suggestion that the subcontractor would be at risk if the court failed to grant an injunction. The Official Referee concluded that it was appropriate to grant an injunction in order to uphold the parties' perceived intentions by entering this form of subcontract.

Perhaps a more interesting question, and one which the Official Referee was not called on to decide, was whether or not an injunction could have been obtained to ensure compliance with an adjudicator's decision ordering money to be paid to a subcontractor. Although the answer is unclear, it has been suggested by some commentators that a subcontractor might additionally be able to rely on a summary judgment application in order to enforce an adjudicator's decision. What is clear is that an adjudicator's decision cannot be registered as a judgment of the High Court under section 66 of the Arbitration Act 1996 for the purposes of enforcement. This case, however, gave some hope to subcontractors, particularly in instances where they were potentially at risk from main contractor insolvency.

Section 9(4) of the Arbitration Act 1996 calls for a mandatory stay of any proceedings brought in the face of an arbitration clause. Thus, existence of the arbitration clause would deprive the court of jurisdiction to decide whether the adjudicator's decision should be enforced (i.e. either by summary judgement or by an injunction); the matter would then have to be arbitrated. The Commercial Court held itself unable to wriggle out of the mandatory stay, a position adopted with some reluctance by the Court of Appeal which divided 2:1 on the point (in *Halki Shipping Corporation* v. *Sopex Oils Ltd* (1998)). The Arbitration Act 1996 may be able to assist with enforcement of an adjudicator's decision. Unless the parties otherwise agree, section 39 permits provisional awards, similar to interim payments in the High Court, and section 48(5) allows an arbitrator to grant injunctions. Thus a claimant, having obtained an adjudicator's decision in their favour, could immediately commence arbitration and seek relief from the arbitrator at an interlocutory stage. Institutional rules have dealt with the problem by 'ring fencing' issues relating to the enforceability of adjudication decisions from the ambit of the arbitration clause.

Chapter 6
Dispute Review Boards and Dispute Advisers

Many construction professionals and lawyers believe that ADR, in the form of non-binding mediation, may impede the proper completion of construction projects if attempted pre-practical or substantial completion. For that reason, in his report *Constructing the Team*[1], Sir Michael Latham considered adjudication attractive. Nevertheless, in the context of large international construction projects, the use of either a dispute review adviser, or more particularly a dispute review board, has become increasingly popular. This is evidenced by FIDIC's publication of provisions for a dispute adjudication board. A dispute review board has not been used much in domestic contracts, however, and with the advent of the Housing Grants, Construction and Regeneration Act 1996 (HGCRA) it is now only likely to have appeal in a limited number of instances. An adviser, or a board, may be invaluable where there is a requirement for swift, efficient, 'binding' (see below) dispute resolution which is seen by a construction team, possibly drawn from different nationalities, to be impartial. There is some flexibility in the role of dispute review boards and dispute advisers; it may vary from simply advising to providing expert determination that is binding. The CIC sees them performing the former of these roles, referring to a dispute adviser as[2]:

> 'A neutral party who simply advises on a problem or potential dispute which requires clarification as to the best method of reaching settlement: sometimes described as Early Settlement Advisor or Early Expert Evaluator.'

Dispute review board

A dispute review board ordinarily consists of three experts (one expert may be used) who are appointed when a contract is awarded but before disputes arise. The members are chosen for their knowledge and technical expertise and their suitability for the range of issues likely to be thrown up by the project in question. The experts become involved in the project from the outset, make regular site visits/inspections and have access to project documentation. Both employer and contractor nominate one member of the dispute review board, with the third member either agreed by both parties or, in the absence of agreement, nominated by an appointing body. Ordinarily the members of the dispute review board will be copied in on all the contract documents, progress reports and other documents which are relevant to the progress of the works.

Once a dispute arises during the construction of the works, which cannot be resolved at site level or by the conventional mechanisms in the contract (e.g. an engineer's decision), the dispute will be referred to the dispute review board. The board will consider the issues raised and, following procedures which can be extremely flexible and tailored to serve the interests of the parties, will provide recommendations to employer and contractor. These recommendations can be open to acceptance or rejection by the parties or have binding effect at the expiration of a prescribed number of days. Alternatively, the parties can agree that the recommendations of the dispute review board remain binding, after an initial opportunity to reject them, until substantial or practical completion of the project has been achieved, at which stage they can be opened up through the medium of arbitration.

The use of dispute review boards is relatively new, with the first such board apparently used on the Eisenhower Tunnel, Second Bore in Colorado, from 1975 to 1979. Since that date boards have been employed on more than $3 billion worth of construction in the USA by a wide range of employers. These include the Alaska Power Authority, California Department of Transportation[3], Massachusetts Water Resources Authority, US Army Corps of Engineers and Washington State Department of Transportation. There is obviously a cost associated with the maintenance and running of a dispute review board. It would appear from US data that a board can cost from $1,000 to $2,000 per member, per one-day meeting. This depends on the location of the board members and the distance they need to travel to the project, the level of their involvement and their charge-out costs. The total cost of a dispute review board over the life of a contract ranges from 0.04 to 0.51% of the total contract cost[4]. Based on ten early contracts, of which the highest percentage related predictably to the least costly contract, the statistic is somewhat unsatisfactory. The average per project has been calculated at 0.17%[5]. It may well be that as contractors become more confident in the capacity of a dispute review board to contain, diffuse and resolve on-site conflict, the costs of maintaining a board will be compensated for, at least in part, by contractors eventually submitting lower tender prices which do not need to build in costings for future expensive litigation or arbitration; although, being realistic this is more theoretical than practical except on very large specialist type work.

The procedures to be adopted by any dispute review board are questions for the parties to decide in conjunction with the potential appointees. However, once a dispute has arisen during construction work on site, which cannot be otherwise resolved, one possible model to follow would be:

• The aggrieved party sets out their position to the board, either by way of an oral statement, or an oral statement based on a previously submitted written document.

- The other party sets out their opening position.
- The parties provide more detailed presentations, referring to particular items of evidence that they consider crucial. There is limited scope for cross-examination by the opposing party and by the members of the board to clarify issues and achieve a prompt understanding of the problem.
- Both parties make their concluding remarks.
- The board retires to consider its views and make recommendations, preferably with a prescribed timetable. The period may be from four days to a week depending on the need to hold a hearing[6]. Ordinarily, once the board's recommendations have been made known they will be open to instant rejection or acceptance by one or other of the parties but, in the absence of a formal objection being made during a prescribed period, the recommendation will become final and binding upon the parties. On most occasions, the board will strive for unanimity in its decision but, with a three-man board, a majority finding may suffice.

The Construction Disputes Resolution Group (CDRG), based on experienced international engineers, enthusiastically promoted dispute review boards. The CDRG produced a booklet[7] that contains useful procedures for mediation, dispute review boards, dispute resolution advisers and cost effective arbitration. With the kind permission of the CDRG, its dispute settlement clauses and guidelines for dispute review boards are reproduced below. These clauses and guidelines are indicative of what may be required.

Dispute settlement clauses

'D.1   Within thirty days of reaching agreement to enter into this contract the parties shall appoint a Dispute Review Board for the purpose of minimising the time and effort necessary to resolve potential claims and disputes which may arise during the progress of the works.

D.2   The Board shall consist of one member selected by the Employer, one by the Contractor and a third selected by the first two. No member shall be selected who is not acceptable to both parties. If there is disagreement as to the selection of any member, then after a further thirty days, the selection shall, on the application by either party, be made from persons nominated by the Construction Disputes Resolution Group.

D.3   The Board shall have discretion as to the conduct of its enquiries. The Board shall visit the site of the works at regular intervals, determined in consultation with the parties but will respond to any urgent request to intervene.

D.4   A Board member may resign from the Board. A replacement member shall be appointed in the same manner as provided for in Clause 2.

D.5 The fees and expenses of all three members of the Board shall be shared equally by the Employer and the Contractor. The Contractor shall make all payments to Board members within thirty days of the presentation of invoices and shall be reimbursed by the Employer.

D.6 The Employer will provide for each Board member one copy of all contract documents, progress reports and other documents pertinent to the activities of the Board. The Contractor shall provide any further documents in its possession which are equally pertinent.

D.7 The Employer will provide all the needs of the Board to enable it to carry out its site meetings function including local travel arrangements, accommodation, conference facilities, secretarial and copying services.

D.8 If the Contractor objects to any act, decision or directive of the Employer, he shall state in writing to the Employer, clearly and in detail, the basis of the objection. The Employer shall respond in writing within thirty days of its receipt.

D.9 Whilst waiting for the Employer's response and throughout the disputes resolution process, the Contractor shall continue to perform the work and conform to the decision or directive.

D.10 If the Contractor is dissatisfied with the Employer's response and the parties cannot amicably settle their difference, either party may appeal to the Board.

D.11 Each party shall supply to the Board complete documentation of its position in relation to the dispute.

D.12 After the Board has considered the matter before it and has given to each party, in writing, its recommendations, each party shall either accept or reject the recommendations by notifying the other party and the Board within 70 days of their receipt. Failure to submit such notification shall be construed as an acceptance of the recommendations.

D.13 Any dispute between the parties which cannot be resolved by negotiation and the assistance of the Board may be settled by arbitration in accordance with the provisions of this contract. All records of the Board shall be admissible as evidence in the arbitration.

D.14 The duties of the Board may be terminated at any time only by the joint decision of the Employer and the Contractor and upon thirty days' written notice to all Board members.'

Guidelines

'1. A dispute arises when a claim or assertion made by one party is rejected by the other and that rejection is not accepted. In order that the relationships between the parties are not soured, the objecting party should immediately submit its appeal to the Dispute Review Board.

2. Normally there will be no great urgency to have the dispute settled but sometimes it may be essential to seek the assistance of the Board as early as possible in order that the matters which led to the dispute are not lost – for example the examination of the condition of a foundation for the concrete sub-structure before the condition is no longer visible due to the pouring of concrete.

3. The Board can be expected to visit the site regularly, typically every three or four months. Each succeeding visit should be fixed while the Board is on site for the previous visit. All fresh relevant documents, particularly periodic progress reports, should be in the hands of members of the Board at least fourteen days before their visit.

4. If an appeal has been submitted to the Board, the parties may be asked for further written documentation and argument to be sent to each member of the Board prior to the next site visit.

5. Each meeting will consist of a round-table discussion and an inspection of the works in order that the Board members may supplement the written information which they have been given. Consideration of each dispute would follow general discussion on the progress of the works, the problems encountered and overcome. The third member of the Board will generally act as Chairman for the main purpose of controlling the meeting.

6. The objecting party would state his case first, followed by the other party. Each would be given the chance of rebuttal and the Board members would ask questions, seek clarification and ask for further data until satisfied that they have all the evidence needed for their deliberations.

7. The Board meets in private and will endeavour to reach a unanimous decision. If the decision is not unanimous, the dissenting member may submit a minority report. The decision with reasons and recommendations are given in writing to the parties. They are not binding but the report (and the minority report if any) will be admissible evidence in any subsequent arbitration.

8. It is desirable not to adopt any rules for the functioning of the Board. The entire process should be flexible and informal. The Board will attend the site for its regular visits even if no disputes have been referred to it.'

Dispute adviser

On occasions, a project does not merit the appointment of a full board and a single dispute adviser may be appropriate. The CDRG also has specimen clauses and guidelines for such an appointment. These are set out below:

Dispute Settlement Clauses

'A.1   Within fifteen days of reaching agreement to enter into this contract the parties shall appoint a Dispute Resolution Adviser (DRA) to minimise the time and effort necessary to resolve potential claims and disputes which may arise during the progress of the works.

A.2   The DRA shall be selected jointly by the parties. If there is a disagreement as to his selection, then, after a further fifteen days, the selection shall, on the application of either party, be made by the Construction Disputes Resolution Group. The appointment of the DRA and his remuneration shall be agreed jointly by the parties.

A.3   The parties shall meet with the DRA within twenty-one days of his appointment to determine the frequency of visits of the DRA and the procedure to be adopted for ensuring his prompt entry into his duties on the request of one or both of the parties.

A.4   The DRA may resign on giving the parties thirty days' notice or his duties may be terminated by notice from the parties jointly. A replacement member shall be appointed in the same manner as provided for in 2 above.

A.5   The fees and expenses shall be shared equally by the parties. Payments shall be made in full for services rendered, the incidentals thereto and all reasonable expenses within thirty days of the presentation of invoices.

A.6   The parties shall provide all the needs of the DRA to enable him to carry out his functions including travel arrangements, accommodation, secretarial and copying services.

A.7   A dispute shall be deemed to arise:
       (a)   when the parties so agree; or
       (b)   when a claim or assertion made by one party is rejected by the other and that rejection is not accepted.

A.8   A party shall request the assistance of the DRA within seven days of a dispute arising but may decide to delay discussion of the issue until the next regular visit of the DRA. If the discussion of the matter is to be so delayed, the parties should attempt to record at the time those matters affecting the issue on which they can agree.

A.9   The parties shall co-operate fully with the DRA in seeking a means of determining their dispute which is acceptable to both parties. Should the matter be within the competence of the DRA and his intervention in the settlement procedure acceptable to both parties, the DRA may act as mediator and shall have full discretion as to the procedure to be adopted.

A.10   If the DRA considers that a settlement could best be achieved by the intervention of an independent third party, he may so advise the parties and seek their agreement to the appointment of a person agreed by the parties or, in the absence of agreement, nominated by the Construction Disputes Resolution Group.

A.11 All disputes and differences not resolved within 60 days of a dispute arising shall be referred to arbitration and final decision by a person agreed between the parties or, failing agreement within fourteen days after either party has given to the other a written request to concur in the appointment of an arbitrator, a person nominated on the request of either party by the Construction Disputes Resolution Group.'

Guidelines

'1. The DRA procedure not only assumes the good faith and trust of the parties but also provides for the means of seeking a settlement to be chosen by the parties themselves with the advice of the DRA.
2. The DRA will meet the parties within twenty-one days of his appointment to agree the frequency of his visits and the action he and the parties are to take when a dispute arises. The Dispute Resolution Adviser will agree with the parties the documentation with which he is to be provided in order that he may be adequately informed at each stage of the work to enable him to carry out his duties.
3. A dispute arises when a claim or assertion made by one party is rejected by the other and that rejection is not accepted. In order that the relationship between the parties is not soured, the objecting party should immediately request the intervention of the DRA.
4. The objecting party should decide whether the dispute justifies the immediate attention of the DRA or whether its resolution can await the next regular visit. Both the DRA and the other party should be notified of the decision within seven days of the dispute arising.
5. The DRA will visit the site regularly, typically every six to ten weeks. Each visit should be fixed while the DRA is on site for the previous visit. All relevant documents which may be of assistance to him should be in the hands of the DRA at least fourteen days before a visit.
6. The notification to the DRA that a dispute has arisen should be accompanied by a brief description of the claim made by the objecting party. The other party should, at the same time or as soon thereafter as possible, make known to the DRA the reason for the rejection of the claim.
7. The meetings with the DRA will be informal and the discussions will take place with no more than two representatives of each party. The intention of the meeting is to determine the best way of settling the dispute. The discussion may well lead to a settlement. It may appear to the DRA that he can guide the parties to a settlement. This he will proceed to do having first obtained the agreement of the parties to his intervention as a mediator.

8. If mediation by the DRA is not appropriate or not agreed by the parties, the DRA may suggest and recommend other ways whereby the dispute might be resolved. These may include calling in an expert to express an opinion on the issues, immediate mediation or conciliation by a neutral third party with the required expertise or immediate arbitration.
9. With the advice of the DRA the parties choose the next stage in the resolution procedure. If no progress can be made, if mediation, conciliation or arbitration is to be delayed to a later date, the DRA will investigate the matter sufficiently to record the facts for use in the subsequent determination of the dispute.
10. If the Dispute Resolution Adviser is requested to give an opinion on any issue, the opinion will be given to the parties within seven days of the request. Where the parties agree to the submission of their dispute to a neutral third person, the procedure will be at the discretion of that person.'

Refinements to dispute review boards

In the USA, there have been two refinements made to dispute review boards, which improve the working of such boards. The Technical Committee on Contracting Practices of the Underground Technology Council adopted a contract provision in three parts:

(1) Escrow of bid documents to be used for facilitating financial negotiations;
(2) A geotechnical design summary report (GDSR) to establish a base line for differing site conditions; and
(3) The Dispute Review Board.

The bid documents to be placed in escrow include all those showing the costs build-up to the successful tenderer's bid. To be placed in escrow means the documents are clearly identified and sealed to acknowledge the formality of the process. The documents remain the property of the contractor but they are placed in escrow for the duration of the project and can be consulted by employer and contractor, at the request of either, to assist in negotiating price adjustments for variations, unforeseen ground conditions, etc. Ordinarily, the documents will also be available to the members of the dispute review board.

The geotechnical design summary report is a statement of the anticipated site conditions; it provides a clear benchmark for the identification of different site conditions and changed geotechnical requirements. Usually, the geotechnical design summary report is a contract document which contains no terms disclaiming responsibility for any inaccuracy or lack of completeness. It is readily available to the members of the dispute review board for their general assistance.

As a major procurer of construction projects, the methods adopted by the World Bank in regard to dispute resolution are worth considering. For instance, the World Bank *Sample Bidding Documents: Procurement of Works*, incorporate the FIDIC 'Red Book' but subject to a number of amendments. The latter are either mandatory, recommended or optional. The World Bank urges that the provisions under clause 67.3 for arbitration be modified:

'In the case of major projects, IBRD (International Bank for Reconstruction and Development) encourages employers to consider introducing a Dispute Review Board (DRB) into the contractual settlement of disputes procedure. Such a DRB could either replace the Engineer under Clause 67 of the FIDIC General Conditions or it could review the decisions made by the Engineer under Clause 67. In either case, both parties should then have the right to request that the decision of the DRB be finally settled under an agreed settlement of disputes procedure by an arbitral award or a court decision. The bank staff is ready to review draft provisions for the introduction of a DRB into Clause 67.'

In 1996 FIDIC published a supplement to the fourth edition of the FIDIC 'Red Book' which provides additionally for an alternative method of dispute resolution using a Dispute Adjudication Board. This is discussed in detail in Chapter 12 which deals with the FIDIC Conditions of Contract for Works of Civil Engineering Construction. The provisions in FIDIC contain some similarities to those of the CDRG set out above.

Andrew Pike had earlier proposed[8] alternative clauses 67 and 68 for the FIDIC 'Red Book' but in doing so he fairly points out a dispute review board will only be suitable for 'heavy' projects: those costing hundreds or thousands of millions of pounds. He proposes the use of an adjudicator for the less weighty project. He also made proposals for setting up a dispute review board where the contract does not incorporate such provisions, but he cautioned against treating his proposals as usable in all situations. He explains the need for elaboration and length in respect of the following terms[9]:

- 'as the procedures are for expert determination, there is no arbitration law to supply details, and there are no applicable standard rules'; and
- 'while the expert determination is subject to review in later arbitration, the decisions of the DRB or the Adjudicator could be absolutely crucial to one or both of the parties, and therefore it must be ensured that each party has a reasonable opportunity to put its case, and be heard.'

Although dispute review boards will continue to be a significant feature of major and complex international schemes[10], the associated costs generally would be prohibitive for run of the mill construction schemes. The use of a claims review board may be desirable, such as on the El Cajón contract in Honduras, which entailed the construction of a concrete arch[11], or on other major projects, such as the Channel Tunnel or Second Severn Crossing. Schedule 5 to the Concession Agreement between the Secretary of State for Transport and Severn River Crossing plc is set out in Appendix [4] with the kind permission of the Secretary of State for Transport. Since the passing of the HGCRA, projects in the UK generally will see the increased use of adjudication but there are many projects that fall outside the scope of the Act and this is discussed in Chapters 5 and 9. Other dispute resolution procedures are available for projects not embraced by the Act, and dispute review boards and expert determination may be used.

Adjudication was the chosen mechanism, for instance, on the Dartford River Crossing[12], a Build Own Operate and Transfer (BOOT) scheme. BOOT contracts have their own particular problems[13]; for example, they usually lack an independent contract supervisor. Disputes may arise as to the adequacy of design work without any effective contract machinery for their resolution. The Dartford River Crossing project documentation provided for the appointment of an adjudicator experienced in cable stay bridge design and construction and associated contractual disputes. The adjudicator was also to have a full understanding of the concession agreement between Dartford River Crossing Limited and the Department of Transport. The adjudication agreement, both for the construction contract and the concession agreement, was based on rule 20 short procedure of the 1983 ICE arbitration procedure. Legal representation was excluded. The parties were to bear their own costs and share the costs of the adjudicator and any assessors he required in the event that technical questions arose which were outside the strict limits of his expertise. Interestingly, adjudication may still be chosen even though the HGCRA does not apply (although in the above case the Act had not been passed) because it is considered desirable. Furthermore, there is freedom as to the choice of adjudication procedures adopted and these may be different from those laid down in that Act.

Chapter 7
Other Forms of ADR

The mini-trial

A mini-trial is considered by some to be a more formalised method of ADR. It is sometimes referred to as an executive tribunal. In the mini-trial each party presents the issues to senior executives of both parties who are often assisted by a neutral chairman. The parties may be, but are not necessarily, represented by lawyers. The chairman, again not necessarily a lawyer, may advise on the likely outcome of litigation but without any binding authority on the parties. After presentation of the issues, the executives try to negotiate a settlement. If successful, the settlement is often set out in a legally enforceable written document. It is important that the executives involved in the process have the authority to settle the matter.

The mini-trial is not really a trial at all (e.g. the legal rules of evidence are usually dispensed with), but a settlement procedure designed to convert a legal dispute back into a business problem. It aims to bring the businessmen on each side of the fence directly into the resolution process in the hope that compromises can be reached. To date the mini-trial technique has been little used in the UK.

The advantages of a mini-trial are:

- A lengthy hearing is eliminated by its summary conduct.
- Each party's case is professionally presented but without any formal rules of procedure or evidence.
- Those who ultimately decide on whether the dispute should be settled (and, if so, on what terms) have the opportunity to be guided by a person with some degree of prestige and outside objectivity.
- The presentations are made to, and the ultimate decision made by, persons with the requisite authority to commit to settlement the bodies which they represent.
- Senior executives from the organisations in dispute are used, who can act subsequently as negotiators, if necessary. It brings to the process of dispute resolution people who can provide a fresh mind. Their lack of previous involvement in the project is a positive feature in that they have none of the prejudices and the false adoption of positions of those whose performance during the course of the actual contract may have been at fault and who now may wish to conceal previous ineptitude.
- Organisations which may have made a long-term investment in

co-operation and which have a number of complex interlinking business transactions can identify common interests which need to be preserved. This may be of special interest to those wishing to develop 'partnering'.

- The use of senior company personnel on the mini-trial panel means that dispute resolution is managed by those who are usually well acquainted with trade customs and practice and technical matters arising both from the questions in dispute and the general organisation of the particular industry.

The disadvantages of a mini-trial include:

- Possible over simplification of complicated technical and legal issues.
- One party may have no real interest in settling and will simply use the process to string the other along.
- Mini-trials are not appropriate for disputes which turn on the credibility of personnel. The mini-trial, with minimal recourse to the assessment of evidence by cross-examination and the substantial use of written statements to provide direct testimony, is ill suited for the verification of the accuracy of what is stated or the assessment of the veracity of particular individuals.
- As with other forms of non-binding ADR, the process is likely to fail whenever a particular party has no real desire to settle and the procedural advantages of the courts favour one of the parties.
- The level of senior management time which needs to be invested in the mini-trial may not render the process cost effective for the smaller dispute.

Although many of the advantages and disadvantages are discrete to the mini-trial, some can equally apply to other forms of ADR.

The first recognised mini-trial is said to have been used between TRW and Telecredit Inc.[1] to resolve a dispute concerning infringement of a computer terminal patent in 1977. In this particular case, the mini-trial was not used early in the dispute resolution process. The parties had already reached the stage of discovery and major negotiations had occurred. Apparently each of the parties had spent in excess of $500,000. Faced with the enormous costs of continued litigation and the potential futility of such a process to the loser, the parties' representatives decided the following:

- No further discovery would occur except that which was necessary for the mini-trial. They agreed to take summary witness statements, which could be amplified, if necessary, at a later date.
- The period for discovery and preparation for the mini-trial would be no longer than 6 weeks, with any dispute about discovery which occurred during that period submitted to the neutral adviser for his non-binding advice.
- The parties agreed to make the appointment of their mini-trial chairman from a range of possible candidates which included a former Supreme

Court Justice and a former Deputy Attorney General of the USA. In the end the appointee was known for his expertise in patent law.

- The parties agreed to share the fees and expenses of the neutral chairman with the fees to be assessed against the eventual losing party if settlement was not reached and a trial resulted.
- Agreements were reached about the material to be submitted to the neutral chairman before the mini-trial.
- All exhibits relied on by the parties were to be submitted in advance to the other party and also to the neutral chairman.
- The parties were to present written opening statements in advance.
- The chairman could submit written questions to the parties' technical experts in advance of the mini-trial.
- Rules of evidence were to be discarded although each party was expected to act in good faith.
- There was to be unlimited scope to decide the material to be placed before the mini-trial panel. The chairman could ask clarifying questions but was not permitted to curtail particular lines of presentation.
- The mini-trial procedure was to be outside the litigation system and no applications to the courts were possible. What was prepared for the mini-trial in written format was not to be admissible at any subsequent trial.
- The chairman could not participate in any subsequent proceedings.
- If either party attempted to make applications to the courts or to call the neutral chairman as a witness, that party would be required to pay the entire costs of the neutral's fees and expenses for the mini-trial irrespective of the outcome of the trial.
- Each side was granted equal time to put its argument and to make a speech in rebuttal of what was submitted by the opposing party. Time was apportioned as follows:

  - Opening presentation by Telecredit – 4 hours
  - Rebuttal statement by TRW – $1\frac{1}{2}$ hours
  - Further rebuttal statement by Telecredit – $\frac{1}{2}$ hour
  - Questions and answers exchanged – 1 hour
  - TRW – 4 hours to present its case
  - Telecredit rebuttal statement – $1\frac{1}{2}$ hours
  - Further rebuttal statement by TRW – $\frac{1}{2}$ hour
  - Question and answer session – 1 hour

- The presentations were mostly made by lawyers but once the mini-trial was at an end the lawyers played no part in the settlement talks between senior management.
- If senior management were unable to negotiate a suitable settlement, the chairman was empowered to provide an advisory opinion on the possible outcome of litigation. The chairman's views were non-binding.

Apparently, senior management only took 30 minutes to reach a settlement following the mini-trial procedure.

Another application of the mini-trial procedure was in a product liability dispute between TRW and Automatic Radio. Automatic Radio brought proceedings against TRW alleging that a circuit board assembly purchased from TRW malfunctioned, thereby causing radios to fail within a few months of purchase. Automatic Radio had sales of $15,000,000–$18,000,000 in the car radio market. It alleged that it was being squeezed out of the market as a result of the errors. The proposed litigation would have turned on complicated technical issues in relation to the design of circuit boards and workmanship. Also, if successful, Automatic Radio would have needed to have damages assessed. One major concern of the parties was to ensure the confidentiality of the process and to avoid the automatic disclosure of information imparted to the mini-trial panel in subsequent litigation. Since a mini-trial is simply one form of settlement negotiation, there should not be any major legal problems provided both parties act in good faith. TRW and Automatic Radio sensibly addressed the question in the document setting up the mini-trial. They included the following confidentiality clause:

'1.   No transcript or recording should be made of this proceeding. All aspects of this advisory proceeding, including without limitation, any written material prepared or oral presentations made between or among the parties and/or the advisor for the purposes of this proceeding are confidential to all persons, including the court, as inadmissible in evidence, whether or not for purposes of impeachment, in the pending civil action or in any litigation which directly or indirectly involves the parties. However, evidence, which would be otherwise admissible if this advisory proceeding did not take place, shall not be rendered inadmissible by its presentation at the advisory proceeding. The advisor will be instructed to treat the subject matter of this proceeding as confidential and refrain from disclosing any of the information exchanged to third parties. The advisor is disqualified as a witness, consultant or expert for either party in this and in any other dispute between the parties. His advisory response, if any, is inadmissible for all purposes in this or in any other dispute involving the parties.'

The mini-trial agreement for the earlier mini-trial between TRW and Telecredit had also contained a confidentiality clause:

'Any violation of these rules by either party will seriously prejudice the opposing party and will be prima facie grounds for a mistrial or disqualification motion.'

A further leading example in the USA of the use of the mini-trial was the multi-million dollar NASA dispute in 1982, again involving TRW, as well as Space Communications Co. (Spacecom). Pre-trial proceedings had taken place over two years. There was concern that a special tracking and data relay satellite system would not be launched on time. The parties agreed to a mini-trial, met and successfully resolved that dispute and other disputes between them in the space of one week. The mini-trial occupied just one day. One significant feature of the NASA mini-trial was that it involved a Government agency which might or might not have had money to pay the costs of the neutral adviser and which was also subject to public auditing considerations. Those problems were overcome and the parties proceeded on the basis of written statements which were exchanged. The mini-trial was held before senior executives. These were the directors of the Goddard Space Flight Center, NASA's associate administrator for tracking data systems, with the president of Spacecom and a divisional vice-president of TRW. No witnesses were called and only the four senior executives asked questions.

The mini-trial technique was used in a dispute between Texaco and Borden. In May 1980 Borden Inc. filed a $200m anti trust suit against Texaco Inc. over a natural gas contract in Louisiana. The Texaco lawyers produced over 300,000 documents in the course of discovery. A preliminary trial was set simply to interpret the contract. A few weeks before the preliminary trial date, counsel for Borden suggested to Texaco's in-house counsel the use of the mini-trial procedure. The proposal was that each lawyer be allowed one hour to present his case in front of the executive vice-presidents of each company. Each party was represented by technical advisers assisted by a third-party neutral. The issues were resolved over a two-week period following the information exchange. The parties negotiated a new gas supply contract that had not been in issue in the original case.

Similar techniques were used in a dispute between Wisconsin Electric Power Company and American Can. American Can sued for $41m for breach of contract relating to industrial waste sold to Wisconsin Electric as boiler fuel. Wisconsin Electric counterclaimed for $20m for the excess cost of burning the waste. The technical issues were obviously complex. A judge was selected as a third-party neutral and there were three days for case submissions. The judge gave his views on the likely outcome of the dispute. Settlement was achieved after a three month period. It has been estimated that the parties saved at least 75 days of trial, many months of protracted discovery and inspection of documents and even more substantial advisers' fees.

Other notable uses of the mini-trial[2] have been in disputes involving federal contracts. Yarn quotes the example of the US Army Corps of Engineers' first mini-trial, which concerned an acceleration claim pending before the Armed Services Board of Contract Appeals (ASBCA). The $630,000 claim was apparently settled in three days for $380,000. The Corps' second mini-trial dealt with a $55.6m claim involving changed site conditions in the construction of the Tennessee Tombigee Waterway. This mini-trial lasted three days, was followed by another one-day mini-trial and produced a settlement of $17.2m. A further example, cited by Yarn, is that of a mini-trial at the Atlanta office of the American Arbitration Association. This involved a $6m claim and counter-claim arising out of the construction of a paper manufacturing plant. The mini-trial panel, composed of a retired federal judge and two senior executives, combined a two-day proceeding with subsequent negotiations over several months. This led to ultimate settlement.

One of the great problems with all forms of ADR, including the mini-trial, is when the process should be initiated. As the mini-trial is a non-binding form of dispute resolution its value will ordinarily be assessed at some stage prior to formal litigation or arbitration. However, the true value of the mini-trial may not be apparent until the precise nature of the dispute is known and it is therefore possible that the mini-trial will be most effective once the dispute has arisen, formal litigation commenced and pleadings exchanged. This creates a problem; once formal litigation or arbitration has been commenced, it is difficult to deviate (a process that the lawyers do not assist) towards a process which is non-adversarial and based on the long-term benefits of compromise. It may well be impossible for the parties themselves to take the giant step necessary to initiate the mini-trial procedure, in which event it may be more appropriate for the party supportive of ADR to use the services of one of the ADR organisations, such as CEDR or the CIArb, to moot the possibilities offered by the mini-trial in the mind of the opposing party.

As there is little experience in the UK of the use of the mini-trial, it is difficult to have any hard and fast ideas drawn from domestic data about the composition of the panel. It is therefore attractive to look to the USA which does have experience of the use of the mini-trial procedure. It is no good choosing the executives simply on the basis of experience. They need to be good listeners, quick-witted, prepared to make concessions when they see their own side's case exposed warts and all, and accomplished in the putting of fact finding questions, as opposed to the making of speeches. They must also have sufficient authority within their organisation to make binding decisions and other necessary compromises.

The choice of the neutral chairman poses different questions. The tendency may be to appoint a legal personality, who will have organisational skills, keep the process moving efficiently and have experience of the area of law in question. This is particularly important if the third party neutral is to advise the parties on the potential outcome of litigation. The legal chairman may be disadvantaged if the parties wish to discuss what the best settlement option would be. The American experience is that parties have shown a preference for retired judges as the third party chairmen in mini-trials.

Although it is on occasions better for the parties to devise their own mini-trial procedure, in the USA both the Center for Public Resources and the American Arbitration Association have published their own mini-trial procedures. Similarly, in the UK the Chartered Institute of Arbitrators has had a mini-trial procedure[3] since 1990. The Chartered Institute of Arbitrators guidelines indicate:

'Most bona fide disputes between reputable parties are capable of settlement in a manner that is business orientated, thus avoiding or at least curtailing legal expenditure, loss of executive time and the deterioration of valuable business relationships.

Such a settlement can be facilitated by the use of a structured procedure which ensures that authorised management representatives are presented with the facts, viewed from both sides, and can then enter into negotiations under the guidance of a neutral adviser experienced in conciliation, mediation and arbitration techniques.'

Following what is described as the explanatory notes, there are five rules:

(1) the procedure
(2) exchange of information
(3) the formal meeting
(4) negotiations
(5) costs.

Helpfully the pamphlet concludes with a draft agreement for parties embarking on a mini-trial.

In the USA, the Center for Public Resources has, in its model ADR Procedures[4], devised a mini-trial programme which is somewhat more sophisticated[5]. The model mini-trial procedure comprises both the rules and a written agreement for the commencement of a mini-trial proceeding. The rules cover the composition of the mini-trial panel (rule 2), the appointment of the neutral adviser (rule 3), discovery of documents (rule 4), brief and exhibits submitted to the neutral adviser (rule 5), the exchange of information between the parties (rule 6), negotiations between the management representatives (rule 7), confidentiality (rule 8) and court proceedings (rule 9).

Some of the rules merit more detailed consideration. For instance, rule 4.1 emphasises that there must be an element of 'cards on the table' from the parties. However, excessive discovery, as found in litigation, is to be avoided. It is therefore the parties' responsibility to endeavour in good faith to agree on appropriate and necessary discovery. Rule 4.2 makes it clear that the parties may, if mini-trial is unsuccessful, revert to traditional discovery in any subsequent litigation. Rule 6 emphasises that each party has the opportunity to present its best case, with the other side being entitled to make a rebuttal. The order and length of presentations and rebuttals is a matter for agreement or, in the absence of agreement, to be decided by the neutral adviser. Flexibility is permitted in the format that presentations and rebuttals should take. The procedure allows for the use of witnesses of fact and expert witnesses. The mini-trial panel is permitted to put questions to those appearing before it to clarify issues, although the strict rules of evidence do not apply. Apart from informal notes, the procedure has no record taken of it. The subsequent negotiations are covered in rule 7 which indicates (rule 7.3) that an appropriate written agreement is signed as soon as agreement has been reached.

Rule 8.1 contains a very important provision:

'All offers, promises, conduct and statements, whether oral or written, made in the course of the proceedings by any of the parties, their agents, employees, experts and attorneys, and by the Neutral Adviser are confidential. Such offers, promises, conduct and statements are privileged under any applicable mediation privilege, . . . and are inadmissible and not discoverable for any purpose, including impeachment, in litigation between the parties to the mini trial or other litigation.'

In line with other dispute resolution schemes, rule 8.2 makes it clear that the neutral adviser cannot play any role in subsequent litigation or arbitration. The use of the mini-trial procedure also acts as a block on the commencement of litigation until the process has been exhausted. If there is already litigation in existence between the parties, when the mini-trial commences, the parties may either ask the court to stay proceedings pending conclusion of the mini-trial proceedings, or request the court to enter an order protecting the confidentiality of the mini-trial and barring any collateral use by the parties of any element of the mini-trial in any pending or future litigation. However, equally importantly, the grant of such a stay and protective order are not conditions for the continuation of any mini-trial proceedings. The procedure comes to an end under rule 10 if the parties fail to execute a written settlement agreement on or before the thirtieth day following conclusion of the information exchanged, subject to agreed extensions, or if either party serves on the other and on the neutral adviser a written notice withdrawing from the proceedings. As with most other dispute resolution schemes, the neutral adviser cannot be sued. He is protected by rule 11.

The matter of privilege, the use of ADR blocking the commencement of arbitration or litigation and termination of the ADR process are discussed more fully in Chapter 8.

Mediation-arbitration (MedArb)

Some lawyers and construction professionals denigrate mediation and other non-binding ADR techniques simply because they are non-binding and are therefore perceived as a weak response to prompt and efficient dispute resolution. This view is not shared by many American lawyers. In one American survey, 85% of those polled[6] did not believe that to propose ADR was a sign of weakness.

One method of addressing concerns about the non-binding nature of mediation is the hybrid technique of MedArb. Its purpose is to commit the parties, usually through a clause in their contract, to continue the ADR processes in a manner that will ensure resolution of the dispute. Assume the disputants will first attempt to negotiate a settlement. If that fails they will embark on mediation and if no agreement is reached the mediator will change roles and become an arbitrator empowered to impose a binding solution on the parties. Many doubts have been expressed, particularly by lawyers, whether MedArb inevitably compromises the neutral's capacity legally to act in an adjudicative capacity while at the same time undermining the efficacy of the initial mediation where the mediator should seek to create an atmosphere of trust and a willingness to impart confidences to him in the caucus sessions. There are other practical considerations. If the parties agree during the original contract negotiations that they will use a form of MedArb in the resolution of any disputes that arise, it may be difficult to assess when the mediation phase should give way to the arbitral one. Simply to place the responsibility on the mediator to advise the parties when mediation should give way to arbitration is an inadequate response.

The value of MedArb is difficult to assess. The technique has not been greatly used in the UK. Even in the USA, with its far longer track record, information is scant. Douglas Yarn concludes[7]:

'For all practical purposes there have been comparatively few detailed reports of MedArb applications in construction disputes. Therefore, most of the available commentary about these hybrid processes is both anecdotal and speculative.'

In his footnote[8] it is stated:

'... One proponent estimates "[t]here are probably thousands of cases..." However, his [the proponent's] definition is so broad as to include the informal use of any third party engaged to render an opinion, binding or non-binding, after failing to mediate a resolution. C.J. Gnaedinger *Mediation-Arbitration: Keeping Conflicts Out of Court*, The Construction Specifier, 54 (1985).'

If mediation fails, the mediator's subsequent appointment as arbitrator of the same dispute is superficially attractive. When played out like High Court litigation, arbitration is expensive. Most arbitrators charge an hourly rate, often with substantial cancellation fees if the arbitration settles at any stage before a full hearing. Also, once a dispute blows up into a full arbitration, cohorts of lawyers and expert witnesses appear to glide effortlessly onto the stage. Anything that may lessen ultimate costs must seem a good idea to the parties. An arbitrator, already well acquainted with the facts by reason of a recently completed, although unsuccessful, mediation, does not have the same learning curve as an appointee coming fresh to the dispute. Such a neutral will have acquired a greater awareness of the dispute than would a conventional arbitrator and is likely to have revealed some of his own impressions as to the weaknesses and strengths of a party's case during the caucus sessions. This might assist parties more easily to draw conclusions as to how a mediator would make a final award in arbitration and, with the issues more clearly delineated, encourage them to deploy simpler arbitration procedures than they would in a conventional arbitration where the arbitrator has to be educated as to the nuances of each party's case. It may be the mediation phase has resolved most, but not all, issues and it is appropriate to permit the neutral to reach a binding decision on the outstanding ones. The realisation that the mediator is subsequently authorised to adjudicate, if necessary, may make MedArb more effective at producing a negotiated settlement than mediation alone.

Merely to follow a clarion call that MedArb is attractive is too superficial. Mediation, unlike litigation or arbitration, succeeds because it is based on communication and trust. A mediator is not constrained to accept one party's case at the expense of rejecting the other party's. The mediation process is designed to release the parties from positional bargaining, frequently articulated as the law's promotion of the establishment of rights, and to look for solutions by a re-focus on interests. With the respect and confidence of both parties, the mediator can listen to the parties communicating confidentially their real positions in the dispute and what they honestly want to achieve. A good mediator will assist the parties in looking for and achieving solutions by identifying what a party really wants. Mediation, founded on a lack of coercion, allows the parties to agree without judicial imposition. The process straddles the law by providing flexible solutions in the form of trade-offs which the law is unable to provide.

In choosing to ignore that mediation is consensual and extra judicial, whereas arbitration is confrontational and part of the legal process, perhaps the hybrid MedArb, although well intentioned, is seriously flawed. Because of pressures on his time a busy arbitrator might coerce the parties during the mediation stage into a settlement which the parties might not desire. Again, knowing that the mediator might subsequently act as their arbitrator, the parties may be encouraged to be less forthcoming. A lazy or inexperienced neutral might cause problems and be inclined to move prematurely to the arbitration phase whenever there was an apparent impasse in the mediation. Influenced by the information released in the mediation, the mediator's subsequent award, when sitting as arbitrator, might owe more to knowledge gained during the mediation than to that admitted in the arbitration phase under the rules of evidence. Regardless of any protestations to the contrary, it is difficult, if not impossible, for any neutral to disregard the parties' confidential compromise positions for settlement expressed during the mediation phase and turn round and make a non-compromising award based on legally assessed rights incorporated perhaps in a reasoned award.

Neither practical nor legal difficulties have prevented the American Society of Forensic Engineers proposing a MedArb clause in the following terms:

'Any and all disputes that arise out of or relate to this agreement, or the performance or the breach thereof shall be subject first to mediation in good faith by the parties and administered by the American Arbitration Association under its Construction Industry Mediation Rules, before resorting to arbitration. Thereafter, any remaining unresolved controversy or claim arising out of or relating to this agreement, or the performance or breach thereof, shall be settled by arbitration administered by the American Arbitration Association under its Construction Industry Arbitration Rules, and judgment on the award rendered by the arbitrator may be entered in any court having jurisdiction thereof. The sole arbitrator shall be the same person as the mediator who is selected under the applicable mediation rules.'

In the USA a number of modifications have sought to address the difficulties posed by MedArb in its basic form where a single neutral is used. Under the first, Med-then-Arb, the mediator and subsequent arbitrator are different individuals. The arbitrator, who is not privy to the mediation phase, will not be influenced by the discussions and materials relating to the previous unsuccessful mediation. Although it solves the problem of having the same individual, it adds time and cost to the dispute resolution process. The appointee as arbitrator has the same learning curve as any other arbitrator. In an attempt to reduce the additional cost the arbitrator may be pre-selected at the commencement of the mediation and at least sit in during the open session presentations to the mediator. The downside obviously is that the parties will have to pay for the arbitrator's time spent in the mediation phase. Although the technique may have some merit where the parties are reasonably confident that not all the issues will be resolved in the mediation, ultimately it is unattractive because it may cause the mediation joint sessions to be conducted in a more adversarial manner simply to impress the prospective arbitrator.

Another modified form of MedArb is one where during the course of the mediation one party may ask the mediator to decide the remaining issues sitting as an arbitrator, provided the other parties concur. If any party objects, arbitration is conducted before a separate neutral. Proponents suggest it encourages the natural tendency of many parties in mediation to ask the mediator to resolve the intractable issues. It is unlikely of course that a disputant who has already done badly in caucus will feel comfortable authorising the mediator to arbitrate. However, the American Society of Forensic Engineers has adopted a standard clause for this type of dispute resolution process.

There are two further major variants of MedArb. Advisory MedArb, where the arbitration award is non-binding, obviously has all the cost and time implications of any other similar process. It might be unsatisfactory for the parties merely to be left with the neutral's opinion of a likely outcome if the unresolved issues were to be settled by means of arbitration. Under concilio-arbitration, which is a form of advisory MedArb, after attempting conciliation, the conciliator produces a draft award, setting out his opinion of the outcome if the dispute were fully litigated. Parties have an opportunity to respond, at which time they can highlight any manifest errors and present further arguments and evidence before the neutral makes a final award. If both parties accept the award, it becomes binding. If the award is rejected it has advisory status. If one party accepts the award and the matter subsequently proceeds to arbitration or litigation, the party rejecting the award may be obliged contractually to pay the whole of both parties' legal costs in the event that the same result is achieved in litigation.

The legal difficulty MedArb has in England and Wales is that it needs to accommodate the requirements of the common law and the Arbitration Act 1996. The revisions to English arbitration law set out in the Arbitration Act 1996 are, with the exception of the provision for appointment of an amiable compositeur under section 46, probably insufficiently radical to take on board the problems posed by MedArb. As practising arbitrators are well aware, under section 23(1–2) of the Arbitration Act 1950 (replaced by 'serious irregularity' under section 68 of the 1996 Act), they risked removal for misconduct, with the result that any award might be set aside if there were grounds to impugn it. Again, under section 24(1) of the Arbitration Act 1950 and Arbitration Act 1996, the court can revoke the authority of an arbitrator because he is not impartial and:

> 'it shall not be a ground for refusing the application that the said party [making the application] at the time when he made the agreement knew, or ought to have known, that the arbitrator, by reason of his relationship towards any other party to the agreement or of his connection with the subject referred, might not be capable of impartiality.'

Section 33(1)(a) of the Arbitration Act 1996 identifies the tribunal's obligations to 'act fairly and impartially as between the parties, giving each party a reasonable opportunity of putting his case and dealing with that of his opponent'. Arbitral 'misconduct' has been widely drawn and, as well as encompassing breaches of the common law doctrine of natural justice, has left many an arbitrator open to attack where, in good faith but misguidedly, he has sought to curtail the procedures, reduce or eliminate the discovery and inspection of documents, or do anything else which he considers at the time conducive to a cost effective and expeditious resolution of the dispute. The Arbitration Act 1996 certainly appears as though it may be more generous in the absence of specific statutory provision for MedArb, but only time will tell.

Lawyers worry that MedArb clauses offend against the principles of natural justice. Many of the decided cases concerning natural justice relate either to civil liberties or are extremely old. In general, they sit unhappily with the aims of modern day commercial arbitration and ignore the fact that commercial parties may at times be prepared to sacrifice some degree of legal and procedural subtlety in the interests of quicker and cheaper dispute resolution. That said, procedural safeguards can never be wholly disregarded. A judge or arbitrator should approach his task with no preconceptions. The application of personal knowledge can remove the capacity to listen attentively to the parties' own views, however misguided. To date the rules of natural justice have developed to service a system of dispute resolution which is adversarial in its conduct, but procedural fairness is essential whenever there is a third party involvement. Neither judge nor arbitrator should express a view before or during the arbitration about the credibility or the veracity of the witness[9]; nor express a view about the merits of the dispute[10]. The test for bias is set relatively low[11] 'was there a real danger of injustice having occurred as a result of the alleged bias' – a judicial relaxation of the 'real likelihood' test[12].

The Latin tag, *audi alteram partem*, describes the other major feature of natural justice. Both parties should have a reasonable and, by implication, equal opportunity to put forward their best case and cross-examine the opponent's witnesses in full knowledge of their case. This explains why an arbitrator should be careful when trying to impose on the parties (with or without their consent) creative approaches to procedure under which the opportunity to cross-examine witnesses is limited and the processes of discovery and inspection of documents curtailed, although the courts are now supporting judges who limit discovery and inspection of documents and boundless oral questioning of witnesses[13].

Today's pragmatic procedural innovation may become tomorrow's ground of appeal for the disgruntled loser. In mediation, confidential information is disclosed by both parties during the mediation phase. If a mediator/arbitrator were to be influenced by such information in reaching his award the losing party would not be fully aware of the case against it and the reference would have been decided other than in accordance with the evidence. Some of the evidence disclosed in the mediation may be for the purposes of the arbitration phase strictly inadmissible. It is invidious for a party not to be able to test such confidential information by cross-examination, particularly if the neutral is about to place reliance on it.

An arbitrator must decide cases judicially by assessing the evidence. Although certain judicial dicta have encouraged arbitrators to adopt an inquisitorial approach, a position endorsed in the Arbitration Act 1996[14], the recognition of the 'amiable compositeur' as part of English arbitral practice has been rather more doubtful, even though now implicitly referred to in section 46 of the Arbitration Act 1996.

Two questions arise in any discussion of MedArb:

(1) Can the rules of natural justice be excluded?
(2) Is a MedArb agreement contrary to public policy, even if both parties originally supported it?

First, it was a development of nineteenth century *laissez faire* principles that commercial contracts left businessmen largely free to contract as they saw fit. To avoid wholly outrageous bargains there has long been the regulatory influence of the courts and more recently statutory control[15]. Again, the courts have tried to be practical about arbitration. In their leading textbook[16], Mustill and Boyd state:

'...where the parties have expressly agreed on a procedure the courts in general will recognise it as a proper, and indeed the only proper, means of conducting the reference.'

The courts' tolerance has extended, on occasions, to permitting legally inadmissible evidence and allowing one party to communicate with a member of the tribunal in the absence of the other. Arbitrators and professionals are well aware of the informality that still attends many trade and quality arbitrations which are conducted on a semi-inquisitorial basis and where procedural rules, if they do exist, are written in a generalised and extremely flexible way.

Ordinarily breaches of natural justice can be waived, although it has been suggested[17] that certain irregularities may be so serious and fundamental in their effect as to amount *ipso facto* to a denial of justice. However, if the parties did have a detailed MedArb clause, a modern commercial court in England and Wales might have some difficulty in ignoring the express wishes of the parties.

For proponents of MedArb clauses, some comfort may be drawn from overseas. For instance, article 10 of the ICC Optional Rules of Conciliation does not preclude the appointment of a former conciliator as arbitrator of the same dispute.

It is easy for lawyers to overstate and exaggerate the difficulties of the MedArb process. Lawyers are by nature conservative and are trained to be cautious. The value of MedArb lies in its supposed fault – there is likely to be an increasingly strong consumer demand for a neutral who can function effectively in both the mediation and subsequent arbitration phases of a dispute. MedArb may assist a party who is in a weaker bargaining position at the mediation stage. A party with a strong case, but an inferior economic position, may feel obliged, having committed time and money to the mediation, to accept a less than adequate solution which could be bettered if it had the appropriate staying power. The mediator's capacity to change from a facilitative to an adjudicative function may help the weaker party by placing the neutral in a position where he has a more intimate knowledge of both parties' cases during the arbitral stage.

Even if the apparent legal difficulties are not fully resolved by this sleight, it may assist if the adjudicative stage is not subject to the Arbitration Act 1996, but instead the adjudicative neutral sits as an expert. Perhaps, however, the best solution is for those who advocate the greater use of the MedArb process to develop a coherent code of conduct for neutrals and the parties using them. By all means allow the neutral to use confidential information revealed to him, provided that it is specifically referred to in any award; and impose a professional duty on the neutral to disqualify himself if he forms the view that for any reason it is not possible for him to act impartially because of the nature of the confidential information disclosed to him during the mediation phase (fraud, dishonesty, wrongdoing, etc.); and above all have clear guidelines as to when the mediation stage is to be deemed concluded and the arbitral phase to begin. Consider whether it is appropriate for such a decision to be made by the neutral; whether it requires a joint decision of the parties; or whether it should simply be the unilateral decision of one of the parties following a defined cooling off period of a specific number of days after the final mediation meeting.

Interestingly, although there is vehement opposition from some English lawyers to the whole idea of MedArb, the same hostility is not found in other jurisdictions. Alan Shilston[18] refers to the Bermuda International Conciliation and Arbitration Bill 1993, since enacted, which contains important conciliation provisions. For purposes of interpretation, conciliation in the Act includes mediation, and 'Conciliation Rules' mean the UNCITRAL Conciliation Rules:

'Where the parties have agreed in writing that a person appointed as a conciliator shall act as an arbitrator, in the event of the conciliation proceedings failing to produce a settlement acceptable to the parties no objection shall be taken to the appointment of such person as an arbitrator, or to his conduct of the arbitration proceedings or to any award, solely on the ground that he had acted previously as conciliator in connection with some or all of the matters referred to arbitration.'

Similarly, there is a Hong Kong initiative that permits mediation to be followed by arbitration, should the parties expressly agree. Shilston[19] states:

'In some jurisdictions, for example Germany, the judges actively attempt to reconcile parties in civil proceedings. It is understood that caucusing does not take place, unlike the US jurisdictions. In the Asian Pacific Rim regional cultural preferences in the matter of third party involvement in dispute settlement lean heavily in the direction of conciliation or mediation. European contractors are active in the region and as a matter of business, particularly the British, should be informed of background state law provisions that exist which allow the possibility of conciliation or mediation to blend with arbitration, should the parties so desire. Regional dispute resolution centres, such as Hong Kong, Bermuda and now Singapore, with Government as a facilitator through statutory enactments, provide settings wherein MedArb could take place.'

A similar position of permitting mediation to be followed by arbitration also exists in Singapore. Shilston[20] refers specifically to the Singapore International Arbitration Act 1994, which contained, among other provisions, the following:

'*Appointment of Conciliator*
16. (1) In any case where an agreement provides for the appointment of a conciliator by a person who is not one of the parties and that person refuses to make the appointment or does not make

it within the time specified in the agreement or, if no time is so specified, within a reasonable time of being requested by any party to the agreement to make the appointment, the Chairman for the time being of the Singapore International Arbitration Centre may, on the application of any party to the agreement, appoint a conciliator who shall have the like powers to act in the conciliation proceedings as if he had been appointed in accordance with the terms of the agreement.

(2) The Chief Justice may if he thinks fit, by notification published in the *Gazette*, appoint any other person to exercise the powers of the Chairman of the Singapore International Arbitration Centre under sub-section (1).

(3) Where an arbitration agreement provides for the appointment of a conciliator and further provides that the person so appointed shall act as an arbitrator in the event of conciliation proceedings failing to produce a settlement acceptable to the parties –

(a) No objection shall be taken to the appointment of such person as an arbitrator, or to his conduct of the arbitral proceedings, solely on the ground that he had acted pre-viously as a conciliator in connection with some or all of the matters referred to arbitration;

(b) If such person declines to act as an arbitrator, any other person appointed as an arbitrator shall not be required first to act as a conciliator unless a contrary intention appears in the arbitration agreement.

(4) Unless a contrary intention appears therein, an agreement which provides for the appointment of a conciliator shall be deemed to contain a provision that in the event of the con-ciliation proceedings failing to produce a settlement acceptable to the parties within 4 months, or such longer period as the parties may agree to, of the date of the appointment of the conciliator or, where he is appointed by name in the agreement, of the receipt by him of written notification of the existence of a dispute, the conciliation proceedings shall thereupon termi-nate.

Power of Arbitrator to act as Conciliator

17. (1) If all parties to any arbitral proceedings consent in writing and for so long as no party has withdrawn his consent in writing, an arbitrator or umpire may act as a conciliator.

(2) An arbitrator or umpire acting as conciliator –

(a) may communicate with the parties to the arbitral pro-ceedings collectively or separately; and

(b) shall treat information obtained by him from a party to the arbitral proceedings as confidential, unless that party otherwise agrees or unless sub-section (3) applies.

(3) Where confidential information is obtained by an arbitrator or umpire from a party to the arbitral proceedings during conciliation proceedings and those proceedings terminate without the parties reaching agreement in settlement of their dispute, the arbitrator or umpire shall before resuming the arbitral proceedings disclose to all other parties to the arbitral proceedings as much of that information as he considers material to the arbitral proceedings.

(4) No objection shall be taken to the conduct of arbitral proceedings by a person solely on the ground that that person had acted previously as a conciliator in accordance with this section.'

Similarly, MedArb is recognised in New South Wales and Alberta to take two further overseas jurisdictions.

Rent-a-judge

The rent-a-judge system has been popular in California for some time, particularly in Los Angeles. It was a reaction to the long periods required to bring actions to trial. It also relies on two particular features of Californian law. Section 638 of the Code of Civil Procedure states that upon agreement of the parties the court can 'appoint a referee to try any or all of the issues in an action or proceedings whether of fact or of law'. Once the judge has designated a particular person to hear the case as referee, that person decides the date and place of a hearing and exercises all the usual powers of a trial judge. The procedure then adopted is exactly the same as that used in Californian trial courts. The court reporter is present and a record made. The referee must make his findings in writing to the court within 20 days of the hearing, stating his findings of fact and conclusions on the interpretation of the law. These findings then become findings of the court and judgment is entered as if the action had been tried in the conventional way. The parties have the right to appeal the referee's decision. The appeal may be either by motion to set aside the referee's report and to ask for a new trial in the lower court or an appeal to the appellate courts on matters of law.

Another feature of Californian law is article 6, section 21 of the constitution. Under this the court may order the case to be tried by a temporary judge who is a member of the state Bar, sworn in and empowered to act until the final determination of the case. A temporary judge has all the powers of an ordinary judge, including the power to commit someone for contempt. The advantage of the constitutional procedure over the statutory one is privacy. The court record will only contain a reference to the appointment of the temporary judge to try the issue in a statement of his decision. There is no requirement for a detailed report which will be made to the court, and thereby become part of the record.

The CEDR has launched its own judicial appraisal scheme[21] as a form of rent-a-judge. It draws on the expertise of senior counsel and former judges in developing the role of ADR in the UK. Members of the scheme include former High Court judges who offer a fast-track approach by which actual or potential litigants can seek an independent assessment of their case by a joint presentation to the senior lawyers on the CEDR panel. Parties can then choose to go on to better informed negotiations or to mediate a settlement. Users of the scheme can agree to treat the assessment as binding. Alternatively, users can use the assessment as a means to assist in the management of cases through the streamlining of issues for future trial or to influence the question of legal costs awards. The English scheme has grown out of models first adopted by the American Judicial and Arbitration and Mediation Services Inc. (JAMS), which was founded in 1979 and has a panel of approximately 200 former American judges. They preside over approximately 10,000 cases annually in 20 or so offices in the USA. JAMS is big business and had a revenue of over $20,000,000 in 1992. Its marketing director attributed its 90% success rate to the former judges' credibility and experience of evaluating facts and legal principles without losing their neutrality.

Other ADR techniques

There are a number of other ADR techniques or refinements of techniques already discussed which are more developed in the USA than in the UK, or which are yet to produce a distinctive English version[22]. These include the following.

The New Jersey medical practice mini-trial

This has been used for medical malpractice cases in New Jersey where liability is admitted. The insurance company and the claimant select a retired judge to listen to a short presentation of the case. with witnesses being heard if necessary. At the end of the presentation, the judge provides the parties with his assessment of quantum. Apparently his advice nearly always results in a settlement, given that under New Jersey procedures the retired judge's decision is admissible in court.

The confidential listener mediation

The parties designate a confidential listener and tell him their maximum and minimum settlement figures. There is agreement that if these numbers overlap, the parties simply settle the case by splitting the difference. If the numbers do not overlap the listener reports this to the parties, and they can try the process again.

The Century City mini-trial

This developed in west Los Angeles and is a modification of the simple mini-trial. It relies on summary arguments by both parties to a three-person panel. The panel consists of the highest ranking officers of the two opposing companies with an impartial presiding officer, usually a retired judge. After the arguments are presented, the two company officers try to agree on settlement. If they cannot, the presiding officer renders an oral or written opinion on what he believes the decision would be if he were to settle the matter as a judge. This may serve to focus minds on the desirability or otherwise of any litigation.

Mini-trial by contract

Instead of simply agreeing to use the mini-trial procedure after a dispute has arisen, parties may agree, as they would agree an arbitration clause, to make provision for a mini-trial as a condition precedent to any litigation. Hancock refers[23] to advice given by an American law professor in a particular case. Each party to the contract was asked to designate an appropriate company officer at subsidiary vice president rank to which any disputes would be referred. If the second-tier corporate executives could not resolve the problem, they would by agreement refer it to their respective chief executive officers for another attempt at amicable resolution. If that failed, the parties could then litigate. The same law professor has also suggested that a provision might be considered which required each party to participate in good faith in a standard mini-trial with some type of monetary penalty against the party which lost the mini-trial, if they commenced legal proceedings and were still unsuccessful.

MEDALOA[TM 24]

This combines mediation with last-offer arbitration. Whenever a dispute involves a question of value or a monetary claim, parties may submit their last demand and last offer to a neutral arbitrator who is simply authorised to select one or the other. The American Arbitration Association proposes two possible clauses for use by parties to achieve this:

> 'The parties hereby submit the following monetary dispute to mediation administered by the American Arbitration Association under its Commercial Mediation Rules and if unable to agree on a settlement amount, they agree to submit their dispute to a neutral person appointed by the AAA who shall select between their final negotiated positions, that selection being binding upon the parties.'

An alternative clause reads:

> 'If a monetary dispute arises out of or relating to this contract, the parties
> agree to first submit it to mediation administered by the American Arbi-
> tration Association under its Commercial Mediation Rules and, if unable
> to agree upon a settlement amount, to submit their dispute to a neutral
> person appointed by the AAA who shall select between their final nego-
> tiated positions, that selection being binding upon the parties.'

The American Arbitration Association suggests that MEDALOA may be the
answer to the problems of MedArb[25].

> 'MEDALOA may be preferable to a similar process called Med/Arb
> which combines mediation with binding arbitration, because MEDALOA
> limits the arbitrator to selecting between the last offers of the parties.
> MEDALOA encourages parties to continue negotiating.'

Chapter 8
Practical and Legal Concerns in Using ADR to Resolve Disputes

Practical concerns

Many clients seem to go to their solicitors' offices in the belief that litigation can resolve all their problems. However, such clients have not made a realistic assessment of their own position. They know what they would like their case to be and what form the facts should take, but are reluctant to look at the reality presented by the facts. In this they are frequently assisted by their lawyers. Most lawyers wish to be helpful and to highlight the positive features in their clients' cases, but it is a bad lawyer who at the same time does not address the client's mind to the weaker or more problematic elements. Those lawyers who do identify weak points are accused of being negative and may be subjected to intense bullying from a client who wishes his lawyer to stand shoulder to shoulder with him without undue dissent.

It is easy to adopt the response, whenever a dispute arises, of 'issue a writ and see what happens'[1]. Such an approach is often supported by the simplistic statement that 'most cases settle'. Indeed, most cases (well in excess of 90%) do settle at some stage prior to trial, but many of those settle only when high legal costs have already been incurred and the scope for creative negotiations has been lessened. By the time settlement is seriously talked about, the substantial fees incurred to lawyers and expert witnesses have become real bones of contention and ones which are accommodated by each of the parties deciding to bear their own costs.

Some will say that it is defeatist to suggest that litigation often resembles pulling the pin out of a hand grenade and then allowing the hand grenade to blow off one's foot. They will point to particular instances where litigation has proved highly successful. However, a situation akin to the former is frequently the case. People do win legal trials but more often than not after a long and bloody battle; conversely legal trials do produce heavy losers. To litigate is to play a lottery: ultimately each party has to possess the capacity to lose. Frequently the use of litigation is justified by stating that the opposing party is wholly unreasonable, someone with whom you cannot negotiate and in any event the prosecution of the claim is a matter of principle. The experienced lawyer advises his clients never to litigate principles, only law. Litigation does have a serious role to play where there are clear legal principles in issue, particularly if they favour one party, but if the dispute is centred on fact, and fact alone, litigation is not the best medium for the resolution of a dispute. Yet, as the primary advisers in any dispute are the lawyers, it needs an experienced, or sufficiently altruistic, lawyer to state that a particular dispute requires a technical assessment rather than risk becoming a playground for legal sophistry.

Traditionally, those who could not resolve their disputes amicably by way of a *sensible* settlement were automatically propelled towards litigation. Fortunately mediation offers the middle way between the unpredictability of litigation and simply giving into a 'bad' settlement in those situations where mediation is a suitable option. In assessing whether or not a case is appropriate for mediation, the following factors may assist a party in making the decision whether to litigate or mediate:

- The parties have and want to maintain a commercial relationship;
- Both parties have a mutual interest in a quick resolution of the dispute;
- Both parties recognise that litigation will provide an unacceptable drain on their managerial time, be expensive, long drawn out and unpredictable;
- Neither party wishes to have the publicity that litigation may bring them;
- The parties have come to understand that mediation may provide them with the best option to have their day in court, a form of catharsis, yet carried out in the most cost effective way possible;
- The parties have already experienced litigation and mediation in other disputes and have learnt the value of mediation and the downside of litigation;
- There may be problems with witness availability or quality and the full intensity of a possible trial is best avoided.

How can a party commence a mediation? The process is not necessarily as simple as issuing a writ because it requires the agreement of both parties to the use of mediation. Mediation is possible in three situations:

(1)   By a contract clause which precludes the use of litigation until the ADR route has been properly explored however questionable the use of such a clause is legally.

(2)   Even if there is no mediation clause in the contract, the parties may agree, on the occurrence of the dispute, that the particular dispute will be referred to mediation rather than to arbitration or litigation in the courts.

(3)   Litigation or arbitration may already have been commenced and the parties subsequently decide that they wish to try and resolve their differences by mediation. A useful time at which to consider the suitability of mediation is at close of pleadings. This will be when the parties have exchanged claim, defence and counterclaim (if any), the reply to the defence and the defence to counterclaim (if any) and completed the task of exchanging and answering requests for further and better information of any pleading. It may not be too late to con-

sider the use of ADR even once discovery (now called disclosure) and inspection have occurred, but before incurring the heavy expense of proofing witnesses, and preparing for and conducting any trial.

The simplest form of mediation to organise is where there is prior agreement via a provision in the contract, and detailed clauses and explanatory notes are available, such as those produced by the CEDR. One form of contractual mediation clause is:

'In the event of any dispute or difference arising out of or in connection with the Contract which the parties cannot resolve by amicable negotiation within [...] weeks from the onset of the dispute, the parties agree prior to any litigation first to try in good faith to settle the dispute or difference by mediation in accordance with the Mediation Rules published by [...]. In the absence of agreement as to the appointment of the mediator, the mediator shall be nominated by the [Centre for Dispute Resolution (CEDR)]. The parties agree to bear equally the administrative costs of the mediation and the mediator's fees. The parties further agree to bear their own fees and costs. The venue for any mediation shall be [London].'

Although parties may choose their own bespoke mediation procedure, either the contract clause or the parties of their own choice may define the procedures to be adopted or refer to a standard mediation procedure of the type found, for instance, in the 1990 Chartered Institute of Arbitrators guidelines for conciliation and mediation (see Chapter 4).

Although ADR, including mediation, appears to have much to commend it as far as clients are concerned, the cry may go up in certain quarters, 'So what's in it for the lawyers?' Many people outside the law do not accept that lawyers have necessarily taken to heart the Bible demand: 'Blessed are the peace-makers: for they shall be called the children of God'[2]. For such people the lawyer's true feelings are nearer to a comment once made (not in jest) by a senior litigation solicitor to one of the authors to the effect that: 'What is the use of a litigation client who is not litigating?'.

Law is, however, a consumer service and the greater the client's awareness of the range of procedures open to their lawyers, the greater the pressure will be from the informed client to produce remedies which are suited to the client's true needs. Busy commercial men should crave solutions rather than sterile and expensive legal battles. Therefore, litigation lawyers cannot ignore the growth of ADR and its attractiveness to a business community that, in the absence of insurance, cannot afford litigation. Similarly, many of the leading insurance companies now endorse ADR and legal aid has been extended to encompass mediation and related techniques. The Technology and Construction Court Solicitors Association (TecSA) is drafting an ADR protocol for use in the Technology and Construction Court.

If lawyers are prepared to view mediation positively, it provides them with the following opportunities:

- They can advise their clients on one of the important factors in the dispute – the client's legal rights;
- They can advise their clients in the choice of a suitable dispute resolution procedure;
- They can assist clients in the preparation of cases for ADR;
- They can represent clients during mediation meetings and mini-trials;
- They can assist clients to prepare and complete appropriate settlement agreements which are legally enforceable;
- They can assist clients to set up contracts which contain more imaginative dispute resolution clauses than the tired formulae that all disputes will be resolved by litigation in the High Court or through arbitration.

Lawyers can also assist with the following:

- Which ADR technique is most suited to the dispute in question;
- Any time limits to be imposed (if none are set in a contractual clause) for the completion of the reference to ADR;
- In the absence of prior agreement, how the mediator should be chosen; should he have particular qualifications;
- What documentation should be prepared and possibly exchanged prior to the mediation sessions;
- Should the mediator be able to make recommendations, either of a non-binding or binding nature;
- Should the mediator provide a written report setting out his recommendations if the mediation fails;
- An analysis of what the disputant's objectives and interests are from resolution of the dispute;
- An analysis of the legal issues, separating them from the factual ones;
- Deciding the method of presentation to be adopted before the mediator; should the lead presenter be technically or legally qualified?
- Analysing whether the dispute has crystallised to a sufficient degree and have the other party's papers been sufficiently disclosed to make possible a sensible assessment of the value of ADR?
- Is it a matter in which the client's professional indemnity insurers are likely to be interested, thereby making notification to them essential?
- A risk assessment of the likely outcome if the matter were to be pursued via litigation or arbitration;
- Are there any general policy considerations, or the requirement for legal precedent, which render litigation in the High Court more advantageous to the client?
- If the dispute were to be litigated or arbitrated in the traditional way, is either party likely to have witness problems – witnesses who are now working overseas or for other employers, witnesses who may be hostile to a former employer, witnesses whose co-operation will be expensive to buy, witnesses whose performance in court is likely to be poor?
- Are the documents in such a mess or lacking in completeness as to render recourse to litigation or arbitration undesirable?

The use of ADR, rather than litigation or arbitration, is not a vehicle for the avoidance of proper case preparation even if the more rapid pace of ADR may lead to more limited (albeit still thorough) preparation than would a full trial. To present a case effectively at a mediation session, to discuss appropriate settlement figures or, as the case may be, trade-offs, all require a full understanding of the issues arising out of the dispute both by the outside advisers and the party's own senior personnel. For the lawyers it is essential to:

- confirm that all necessary information is available for the mediation sessions
- know the facts of the dispute thoroughly from both perspectives, take necessary witness statements to be incorporated into a position paper, collate these and understand them thoroughly
- identify and analyse the important or likely to be contested legal issues
- complete a risk analysis chart of the strengths and weaknesses of the client's case and, as far as possible, the strengths and weaknesses of the other side's case
- determine who should attend the mediation sessions on behalf of the client, ensuring that those who do attend have sufficient authority and are of sufficient standing within the organisation to reach binding agreements, make concessions and engage in necessary trade-offs
- determine, in conjunction with the client's representatives, the best and worst positions on liability and, as far as possible, quantum giving due regard to those issues in the opponent's case which may objectively cause a change of opinion
- develop a negotiation plan which accommodates the offer or demand that the client would ultimately be happy to settle for, following the mediation sessions
- decide, prior to the mediation sessions, if there are any facts that are not to be disclosed to the other party, so as to be prepared to inform the mediator immediately about them in strict confidence
- be aware that there is no value in simply holding on to information for the sake of it, adopting the litigator's favoured stance of 'keeping your powder dry'
- consider with the client whether the opening address in joint sessions with the mediator and the other party will be made by the client or by the lawyers
- produce an analysis of the likely costs pattern if the dispute were to be resolved by litigation or arbitration, including provision for expert witness and counsel's fees
- assess whether there is any value in proceeding via litigation to create a precedent or because important points of contract interpretation arise or issues of public or administrative law need to be resolved
- consider even if judgment is achievable through litigation, how recoverable is that judgment

- assess what is the hidden cost to the client in having key personnel tied up in the preparations for trial, potentially for many months
- ascertain whether there are any tax advantages that may be available to either party if a dispute is resolved quicker through mediation at a particular time rather than years hence through the courts or arbitration.

Legal concerns

Limitation periods

Whenever a would-be claimant is considering how best to resolve his dispute, both he and his lawyers must remember the importance of the limitation periods. If a claim is to be litigated through the courts or through arbitration, that claim must ordinarily be commenced within the statutory period of limitation. The requirement to assert the relevance of the limitation period is the defendant's; the question of limitation may serve as part of his defence to the claim. The Limitation Act 1980 is the primary source of the law on limitation of actions. This provides that:

- An action arising out of a simple contract must be brought within 6 years of the date when the cause of action accrued (i.e. the date the breach occurred) (section 5); and
- Any claim under a contract executed as a deed must be brought within 12 years of the date on which the cause of action accrued (section 8).

The trigger date is the date when the breach of contract occurred. This is often quite difficult to assess in construction claims. If, for instance, the claim is based on defective work it may be extremely difficult to ascertain the date on which particular work was carried out. Therefore, construction contracts often assume that the date of breach is the date of practical or substantial completion. That does not exclude other possible dates. For instance, if the contractor fails to carry out an instruction to make good defects, that may well constitute a further breach of contract on his part. Indeed, the contractor will not finally discharge his obligations until the end of the defects liability or maintenance period. He may therefore be liable for the whole of that period. A further complication in fixing limitation periods results from section 32 of the Limitation Act 1980. This deals with deliberate concealment on the part of the defendant. The commencement of the limitation period is, in such circumstances, suspended until such time as the claimant has either discovered the concealment or is in a position where he could, with reasonable diligence, have discovered it.

Many claims brought are argued by lawyers both as breaches of contract on the part of the defendant and, in the alternative, as giving rise to causes of action in tort (negligence). For a time it was suggested that a claimant who had a detailed contract with the defendant could not at the same time assert rights against the defendant in the law of negligence[3]. However, the courts have re-asserted with some vigour that the existence of a formal contract does not necessarily preclude the existence of concurrent rights in negligence which may, on occasions, be more extensive[4]. Obviously a claimant cannot obtain double recovery, however his claim is argued.

The Limitation Act 1980 affects negligence claims quite differently from contract ones. Section 2 of the Limitation Act 1980 provides that no action founded on tort can be brought more than 6 years from the date on which the cause of action accrued. This is the date when damage was suffered. This may be long after the date on which a breach of contract occurred. This means the claimant has potentially a longer time period in which to bring his claim. The fact that a cause of action in tort arises when damage occurs is a possible source of injustice to complainants. There could be circumstances in which a claimant, through no fault of his own, was unaware that he had in fact suffered damage. The House of Lords decision in *Pirelli General Cable Works Limited* v. *Oscar Faber & Partners* (1983) is a notorious example of this.

Oscar Faber, a well known firm of consulting engineers, designed a chimneystack which was built in 1969. According to expert evidence, cracks developed near the top of the chimney not later than 1970, which would not have been noticed on general routine maintenance. The damage was not discovered until 1977 and a writ was issued in 1978. The House of Lords, while recognising what was at the time the unsatisfactory state of the law, held that the plaintiff's cause of action in tort arose in 1970 and was therefore clearly statute barred. The consequence of the *Pirelli* decision was that parliament enacted the Latent Damage Act 1986.

This Act introduced new sections 14A and 14B into the Limitation Act 1980. These were a material change to the old law. They allowed a plaintiff, who had not commenced proceedings within six years of incurring damage, to commence proceedings within a further three year period from the date on which he first became aware that he had incurred damage, if his failure to become aware of the damage within the primary six year period was reasonable. In addition, to bring some certainty to the defendant's position, parliament also created a long stop period of 15 years, effectively extinguishing the plaintiff's right to claim, calculated from the date of the last alleged act of negligence on the part of the defendant.

The importance of recognising statutory limitation periods cannot be overstated. No defendant is going helpfully to draw a would-be claimant's attention to the need to bring his claim in court or arbitration before it becomes statute barred. Such considerations become all the more critical when discussing ADR solutions. No claimant can afford to become so embroiled in the ADR process as to ignore the time when issue of a protective set of proceedings to avoid a particular claim becoming statute barred may become essential, rather than simply prudent. Conversely, it might benefit a cynical defendant to go through the motions of seeking an ADR solution with a particular claimant, but with no intention of taking the process seriously; simply using it as a stalling tactic in the hope that the claim will either go away through inertia on the part of the claimant or, as an attractive alternative, subsequently be found to be statute barred.

A further complication under certain construction contracts is the existence of a distinctive dispute resolution mechanism which needs to be followed, and the various standard forms provide a number of examples. An obvious example is found in the ICE Conditions of Contract 6th Edition where the timetable laid down in clause 66 needs to be followed carefully, and similarly under JCT 80 and JCT 98 the time available to open up a final certificate is circumscribed.

Although there is a general power under section 27 of the Arbitration Act 1950 (broadly replicated in section 12 of the Arbitration Bill 1996) for the High Court to extend the period during which arbitration proceedings may be commenced (to avoid undue hardship to a claimant) it is unreliable to rely on the exercise of the discretion in construction cases. There are two building cases on the point and a civil engineering one. The building cases both relate to JCT 80. In the first, *McLaughlin & Harvey* v. *P & O Developments Limited* (1991), the Commercial Court held that the period during which arbitration proceedings could be commenced in regard to a final certificate could be extended by the court under section 27 of the Arbitration Act 1950. In the more recent case, *Crown Estate Commissioners* v. *John Mowlem & Co. Limited* (1994) the Court of Appeal held that the decision in *McLaughlin & Harvey* was incorrect and that there was no discretionary power under section 27 of the Arbitration Act 1950 to extend the period during which arbitration proceedings could be commenced. The latter decision does not sit happily with *Christiani and Nielsen Limited* v. *Birmingham City Council* (1995). In this case two issues were before Judge Hicks QC, including: could the three-month period to commence arbitration proceedings under clause 66(3)(a) be extended under section 27 of the Arbitration Act 1950? For reasons relating to the particular case before him, the judge was not obliged to consider the issue in detail. However, he indicated that he would have approved a section 27 extension if necessary. Case law is already demonstrating that the section 12 test under the 1996 Act is more restrictive (*Fox and Widley* v. *Guram and Another* (1998) and certain shipping cases including The Catherine Helen (1998) and *Grimaldi di Compagnia di Navigazione SpA* v. *Sekihyo Line Ltd* (1998)).

Achieving certainty in the ADR process

One of the constant criticisms of ADR is that, even if successful, any agreement is difficult to enforce legally if a party later reneges. Some certainty is possible if, following a successful mediation, the parties set out the terms of their settlement in writing and in a form which is enforceable in the courts without further analysis of the issues. Here lawyers can assist. First, although there has not been a judgment of the courts, the agreement can be written so as to be an enforceable agreement which, if necessary, the courts can be asked to enforce. Such agreements can be based on so-called Tomlin Orders which have been frequently used in litigation where complex terms are agreed. They are named after Tomlin J. although under Part 40 of the CPR they are more prosaically described as consent orders. In a practice note[5], he said that where terms of compromise were agreed with the intention that an action in the courts would be stayed in accordance with the terms scheduled to the order, the order should be worded as follows:

'And, the plaintiff and the defendant having agreed to the terms set forth in the schedule hereto, it is ordered that all further proceedings in this action be stayed, except for the purpose of carrying such terms into effect. Liberty to apply as to carrying such terms into effect.'

In litigation, if the agreed terms are breached, enforcement requires first that the action is restored under the 'liberty to apply' provision with an order then obtained to compel compliance. Second, if that order is itself breached, enforcement is then possible in court. If an ADR settlement has occurred, following the earlier issue of a writ, an unamended Tomlin Order is applicable. If there has been no prior or concurrent litigation, the parties will need to produce a specially drafted agreement which includes the following:

- Identity of the parties;
- Recitals indicating that various disputes have arisen, the disputes have been referred to mediation by ... and that settlement has now been achieved in the terms set out below;
- The terms of settlement;
- A provision to the effect that the agreement shall be deemed to have between the parties the same effect as if the agreement had been reached as a result of or during the course of litigation and that any party finding another party to be in breach of the terms of the agreement is at liberty to apply to a nominated court for necessary orders to ensure compliance by the recalcitrant party with the terms of the agreement.

A second method to assist enforcement has been to state that the agreement shall be deemed to have the same effect as an arbitration award to which the Arbitration Act 1996 applies. Then, if necessary, enforcement of the 'award' could have been be made through section 66 of the Arbitration Act 1996, which states:

'An award on an arbitration agreement may, by leave of the High Court or a judge thereof, be enforced in the same manner as a judgment or order to the same effect, and where leave is so given, judgment may be entered in terms of the award.'

'(1) An award made by the tribunal pursuant to an arbitration agreement may, by leave of the court, be enforced in the same manner as a judgment or order of the court to the same effect.
(2) Where leave is so given, judgement may be entered in terms of the award.'

One cannot be sure how the mediation will end and whether the parties will reach agreement or not. What will happen if there is a failure to agree? Is there to be a recommendation from the mediator? What happens following the termination of an unsuccessful mediation session; is there to be a cooling off period before either party may have recourse to the courts or arbitration? Clearly, these are matters that are best dealt with at the outset of the mediation.

Privilege and witness compellability

The doctrines of privilege and witness compellability may cause considerable problems in the context of ADR. Difficulties are not assisted by the lack of direct judicial guidance. Questions include:

- If a mediation fails and the parties return to the courts or to arbitration, what is the status of documentation prepared and tendered for the purposes of the failed mediation? Is it safe from disclosure in subsequent court or arbitration proceedings?
- Is the mediator at risk of being called to give evidence in subsequent court or arbitration proceedings on behalf of either one or other of the parties?
- Can the mediator remain silent concerning comments made to him during the private caucus sessions?
- Whether the mediation is successful or not, what happens if there is subsequent related litigation and one or other of the parties considers what happened in the previous mediation to be relevant and either seeks specific discovery of mediation documents or attempts to require the attendance of participants in the mediation at the subsequent litigation?

Although parties often state expressly that non-binding mediation is to be carried out between them on a 'without prejudice' basis, even if the process is not expressly stated to be without prejudice it would probably be treated as such by lawyers. Non-binding mediation has much in common with ordinary settlement talks that parties to litigation might attempt. The phrase 'without prejudice' simply means that if settlement talks are unsuccessful, any statements made will be privileged; no reference can be made to them in any subsequent litigation or arbitration proceedings. In *Rush & Tompkins Limited* v. *Greater London Council* (1989), Lord Griffiths said[6]:

'The rule applies to exclude all negotiations genuinely aimed at settlement whether oral or in writing from being given in evidence. A competent solicitor will always head any negotiating correspondence "without prejudice" to make it clear beyond doubt that in the event of the negotiations being unsuccessful they are not to be referred to at the subsequent trial. However, the application of the rule is not dependent upon the use of the phrase "without prejudice" and if it is clear from the surrounding circumstances that the parties were seeking to compromise the action, evidence of the content of those negotiations will, as a general rule, not be admissible at the trial and cannot be used to establish an admission or partial admission.'

The privilege in statements made on a 'without prejudice' basis is the joint one of the parties and extends to their solicitors[7]. It can only be waived with the consent of each of the parties.

Some assistance can be drawn from matrimonial cases which may be subject to conciliation. If negotiations have taken place in the presence of a mediator or a counsellor, with offers and suggestions being relayed to the parties via the neutral, those negotiations are privileged. They have the same protection as if made in correspondence. In a relatively old case, *McTaggart* v. *McTaggart* (1949) an interview had been arranged between spouses in front of a probation officer on a 'without prejudice' basis. The Court of Appeal held that either spouse was entitled to object to evidence of what had been said being admitted at a subsequent trial. However, as the privilege was that of the parties, the probation officer was not able to object if the parties then chose to waive the privilege in the statements made before the probation officer. The judge was bound to admit them as evidence at trial. The modern law, at least in the context of matrimonial proceedings, was stated in *D* v. *NSPCC* (1978)[8]:

'With increasingly facile divorce and the vast rise in the number of broken marriages, with their concomitant penury and demoralisation, it came to be realised, in the words of Buckmill LJ in *Mole* v. *Mole*: "in matrimonial disputes the state is also an interested party; it is more interested in reconciliation than in divorce". This was the public interest that led to the application by analogy of the privilege of "without prejudice" communications to cover communications made in the course of matrimonial conciliation (see *McTaggart* v. *McTaggart*; *Mole* v. *Mole*; *Theodoropoulas* v. *Theodoropoulas*) so indubitably an extension of the law that the text books treat it as a separate category of relevant evidence which may be withheld from the court. It cannot be classed, like traditional "without prejudice" communications, as a "privilege in aid of litigation"...'

A situation could arise in mediation akin to that in *Rush & Tompkins* v. *Greater London Council*. The plaintiff brought an action against two defendants but eventually settled with the first defendant. The second defendant sought disclosure of the 'without prejudice' negotiations between the plaintiff and the first defendant which were obviously relevant to the action between the plaintiff and the second defendant. In reversing the decision of the Court of Appeal, and upholding that of the first instance judge, the House of Lords held that the provisional 'without prejudice' negotiations between the plaintiff and the first defendant were subject to privilege. According to Lord Griffiths, a view concurred in with the other members of the House of Lords, it was untenable to suggest that once negotiations were successful, privilege in the 'without prejudice' correspondence had served its purpose and must be disregarded.

The question of privilege, arising out of a mediation hearing has been addressed in the United States and Australia. The Australian position is summarised in Chapter 3. In the United States the Southern District Court of New York has ruled that documents from an ADR proceeding are protected from discovery and subsequent court proceedings under the attorney–client privilege and work product doctrine. This ruling was made in the case of *North River Insurance Co.* v. *Columbia Casualty Co.* (1995). North River claimed losses incurred for asbestos-related defence costs under an insurance policy with Owens-Corning Fiberglass. Following an ADR procedure between North River and Owens-Corning, North River was required to pay Owens-Corning's defence costs. North River then sued a number of re-insurers, including Columbia Casualty, to recover part of the defence costs. Columbia Casualty argued that such costs were not covered under the original policy between North River and Owens-Corning and were therefore not subject to the re-insurance agreement. Columbia Casualty requested discovery of all of North River's documents relating to its dispute with Owens-Corning. North River requested a protective order from the Southern District Court on the grounds that documents disclosed in the context of an ADR procedure were protected from discovery under the attorney–client privilege and work product doctrine. The court found that Columbia Casualty and North River were not represented by the same counsel, did not share legal expenses, did not pursue a co-ordinated litigation strategy and therefore lacked a common interest.

In April 1996[9] the American state of Pennsylvania brought into effect a law granting statutory privilege to most communications resulting from mediation:

> 'disclosure of mediation communications and mediation documents may not be required or compelled through discovery or any other process. Mediation communications and mediation documents should not be admissible in any action or proceeding, including, but not limited to, a judicial, administrative or arbitration action or proceeding.'

By this law, Pennsylvania aims to encourage mediation and to avoid any risks that the court may subpoena parties or mediators to obtain documents and information. The statutory exemptions are:

- Settlement documents which may be introduced in an action to enforce settlement;
- Any communication or conduct that is relevant evidence in a criminal action;
- Fraudulent communications made during mediations which become relevant evidence in an action to enforce or set aside a mediated agreement as a result of fraud.

In order to protect mediators from the risk of becoming involved in any subsequent litigation or arbitration proceedings, most mediation rules provide that an appointed mediator, who is under contract to the parties, is expressly precluded from being called to give evidence in any subsequent litigation or arbitration. In any event, most mediators take practical measures to ensure that they reduce the possibility of being subjected to document subpoenas or any other court process. More often than not the parties will agree that any documents prepared for and in the course of mediation, including the mediator's notes, will be destroyed at the termination of the proceedings.

As mediation remains, in the UK, a relatively little used, albeit growing, technique, there is relatively little direct authority from the courts relating to the possible problems that it can create. One of the better documented questions is of the enforceability of mediation clauses.

Enforceability of ADR clauses

What happens if one of the parties decides not to play ball?

If ADR is a contractual obligation and an impatient party tries to bypass the contractual machinery and commence immediate litigation, the courts do have inherent powers to stay any action which they consider should not be allowed to continue:

> 'Where parties have agreed the machinery ... for the resolution of disputes, it is not for the court to intervene and replace its own process for the contractual machinery agreed by the parties.'[10]

English authority is scant, save in regard to so-called expert determination, on the question of the extent to which the courts will uphold non court-based procedures and 'lock-out' court applications. The position in both the USA and Australia is different. In *Hooper Bailie Associated Limited* v. *Natcom Group Pty Limited* (1992), both parties to a construction dispute had agreed to conciliation in order to accelerate resolution of their dispute. The Supreme Court of New South Wales held, inter alia, that as the parties had agreed to conciliate, the court had power to order a stay of arbitration until the conclusion of the conciliation procedure. A similar result was achieved in *Elizabeth Bay Developments Pty Limited* v. *Boral Building Services Pty Limited*

(1995), although on the facts of the case the particular mediation agreement was unenforceable because it lacked certainty.

In *Public Authority Superannuation Board* v. *Southern International Developments Corporation Pty Limited and Another* (1990), disputes about any matter relating to the construction of a shopping centre in Sydney were referred to an engineering expert. The contractor made a claim for extra costs arising from variations and initiated the reference of the claim to an engineering expert. The owner objected but the court upheld the clause. The USA courts have upheld alternative dispute resolution clauses, where they were part of the contract between the parties, as a necessary first step prior to any litigation or arbitration. The District Court of Oregon specifically approved in *Haertl Wolff Parker Inc.* v. *Howard S. Wright Construction Co.* (1989) the earlier decision in *Southland Corporation* v. *Keating* (1984) to the effect:

'A contract providing for alternative dispute resolution should be enforced and one party should not be allowed to evade the contract and resort prematurely to the Courts.'

Against the general trend of both USA and Australian decisions, in *Allco Steel (Queensland) Pty Limited* v. *Torres Strait Gold Pty Limited* (1990), the Queensland Supreme Court refused an application to stay litigation commenced in contravention of a conciliation clause. The reasoning was as follows:

'... [the contract] merely provides an agreement to conciliate [as distinct from one to litigate] and as such is severable from the binding agreement in which it is located. In other words, notwithstanding what I perceive to be a clear breach of the obligations to conciliate on the part of the plaintiff, the doctrine that the jurisdiction of the court cannot be ousted dominates any other principle that would require the plaintiff to honour its contractual obligations.

An appeal was made to the inherent jurisdiction of the court to grant a stay, the condition precedent to the accruing of a course of action not having been met, namely bona fide conciliation. In my view, even if such relief was open, this discretionary relief must be refused as it is abundantly clear that the parties have taken up positions which effectively rule out compromise and conciliation...'

In a later case, *AWA Limited* v. *Daniels and Others* (1992), the Supreme Court of New South Wales expressly disapproved *Allco Steel*. Rogers J considered that the commencement of formal litigation without complying with the contractual provisions for mediation was an abuse of the process.

The question of non-binding mediation is more interesting and problematic than binding conciliation or expert determination. Whenever parties embark on this form of ADR they understand that the procedure does not necessarily mean a successful outcome. It is a trite principle of English law (unlike the position in either the USA or Canada) that an agreement to enter into negotiations is unenforceable. For instance, Lord Denning MR stated in *Courtney and Fairburn Limited* v. *Tolaini Brothers (Hotels) Limited* (1975)[11] that such agreements were 'too uncertain to have any binding force'. Australia too has adopted the English approach (*Coal Cliff Collieries Pty Limited* v. *Sijehama Limited* (1992)).

In what some commentators considered to be a retrogressive decision, the House of Lords did not deviate from this stance in *Walford and Others* v. *Miles and Another* (1992). The husband and wife defendants were the owners of a company and property from where a photographic processing business was carried on. In January 1987 negotiations began between the first and second plaintiffs and the first defendant for the sale of the company and the property. On 17 March 1987 the first defendant orally agreed to deal exclusively with the first plaintiff and to terminate any negotiations then being carried on between the defendants and any other potential purchasers. The only condition was that the plaintiffs provide a 'comfort letter' confirming that they had the necessary financial support from their bank to complete any purchase. The condition was complied with. Notwithstanding the 'agreement', the defendant subsequently chose to deal with a third party and completed a sale which excluded the plaintiffs. The plaintiffs sued for breach of contract. The House of Lords held that the agreement of 17 March contained no term as to the duration of the obligation to negotiate with the plaintiffs and made no provision for the defendants to determine the negotiations. Any duty to negotiate in good faith was unworkable and inherently inconsistent with the position of a negotiating party, since parties to negotiations were always at liberty to terminate such negotiations at any time and for any reason. The agreement as between the plaintiffs and the defendants was void for uncertainty. It was simply their agreement to negotiate. In the words of Lord Ackner[12]:

'An agreement to negotiate has no legal content' and '...good faith is inherently repugnant to the adversarial position of the parties when involved in negotiations.'

In some respects, there was a superficial similarity between *Walford* and a later Court of Appeal decision, *Pitt* v. *P.H.H. Asset Management Limited* (1994). The case related to the sale and purchase of a dwelling house. The parties had agreed a 'lock-out' arrangement by which the plaintiff should have the opportunity to exchange contracts within two weeks of receipt of the draft during which period the defendant would refrain from negotiating with third parties. The Court of Appeal upheld the 'lock-out' arrangement in *Pitt* because it was for a finite period and therefore not uncertain. In *Pitt*, Peter Gibson LJ quoted from Lord Ackner's judgment in *Walford* at page 139[13]:

'There is clearly no reason in the English contract law why A, for good consideration, should not achieve an enforceable agreement whereby B agrees for a specified period of time, not to negotiate with anyone except A in relation to the sale of his property.'

and then continued:

'... B by agreeing not to negotiate for this fixed period with a party, locks himself out of such negotiation. He has in no legal sense locked himself into negotiations with A. What A has achieved is an exclusive opportunity, for a fixed period, to try and come to terms with B, an opportunity for which he has, unless he makes his agreement under seal, to give good consideration.'

It is reasonably clear that in the absence of any apparent direct English authority, the courts would have considerable difficulty in holding that a provision for non-binding mediation which set no time limits in which the process was to be carried out, was legally enforceable. By contrast, there is every reason to believe that the courts would be supportive of non-court based adjudicative methods of dispute resolution (however described) which either set clear time limits for a result to be achieved or which led to a binding decision being imposed on the parties, even if it was merely one of an interlocutory nature which was open to subsequent review by the courts.

Chapter 9
Adjudication and the 'Construction' Act

Legislation on construction contracts

During the writing of this book, the consequences of Latham started to emerge. Latham's report *Constructing The Team* has initiated a range of developments touching on many aspects of the construction industry including adjudication.

In May 1995 the Department of the Environment (DoE) published a consultation paper entitled *Fair Construction Contracts* which sought the construction industry's views on a range of matters including adjudication. The UK Government had indicated a willingness to legislate, if general support were to be found within the industry. No general agreement on solutions was forthcoming, except to the idea that legislation may offer a solution to some of the problems encountered by the construction industry. In some ways, it was odd to single out the construction industry. There would seem to be no logical reason to treat one industry differently from another and to do so will depict it in a way that is counterproductive. Notwithstanding the lack of any genuine agreement on the need and nature of legislation, the Government was nudged, if not coerced, into proposing legislation. It seemed that the process had started to roll and that whatever misgivings there might be concerning legislation, it was easier to agree with the idea, shaping and containing its impact, rather than resist its introduction. Subsequently, legislation in the form of the Housing Grants, Construction and Regeneration Bill was proposed.

Part II of the Housing Grants, Construction and Regeneration Bill covered construction contracts that incorporated, among other things, a right to refer disputes to adjudication. The Bill also proposed secondary legislation in the form of the Scheme for Construction Contracts that was to be applied in the absence of appropriate adjudication provisions within the contract. In the words of Robert Jones, Minister of State at the DoE, an 'illustrative' draft of the Scheme was published to show how the primary and secondary legislation would fit together. Although the Bill attracted a good deal of attention from bodies representing all sectors of the industry, the Scheme, because of the detailed and often contentious provisions, caused greater concern. The Scheme provisions would become the meat of the legislation and they would be applied to all construction contracts, thus removing the freedom for parties to make their own suitable adjudication provision within the contract.

The Bill received a great deal of consideration by the main industry groups as it passed through its legislative stages. Greatly amended, its final form remained very uncertain until it became an Act. In fact, new wording was incorporated during July 1996, the month that the Bill became an Act of Parliament.

Housing Grants, Construction and Regeneration Act 1996

Although this Act received royal assent in July 1996, Part II dealing with adjudication would not apply until the 'commencement of this Part'. This was 1 May 1998. The operation of the Act was dependent on the existence of a Scheme for Construction Contracts because a fallback position had to be in place if contracts were defective. The order for the Scheme was made on 6 March 1998 with an 8 week lead-in period. The delay in operation of the Act was primarily to enable a revised Scheme for Construction Contracts to be prepared by the DoE, for there to be proper consultation as required by section 114(2) of the Act, and for the minister to introduce the Scheme by way of regulations.

The delay in the commencement of the implementation of Part II also provided time for the contract authoring bodies to make whatever amendments they felt were necessary to ensure their standard contracts complied with the requirements of the Act. As a result, the Scheme may have limited effect, although many construction contracts are let on non-standard forms. Some of these will not provide for adjudication as statute requires or will misapply the principle of the Act and consequently the Scheme will be applicable. Furthermore, not all authoring bodies had amended their standard contracts to comply with the Act by the commencement date.

The Government indicated its intention to become a 'best practice' client and used this period to amend its own forms of contract. However, not all government departments, notably the Ministry of Defence, are well disposed to amending their procedures. JCT's consideration of the inclusion of adjudication provisions was well advanced at the Scheme consultation stage but the publication of amendments prior to the approval of the Scheme was something that they did not wish to do. The JCT felt the Scheme provided an insight into the application of the Act; hence, contract drafters were nervous about committing themselves.

Although the HGCRA does not apply to contracts entered into prior to commencement of Part II, the parties to existing contracts may adopt adjudication by agreement. Furthermore, there are also exempt categories of work under the Act, and it does not apply to work outside England, Scotland, Wales and Northern Ireland. It may be argued that a contract was entered into prior to commencement of Part 11 and there may be problems with main and subcontracts entered into pre and post the Act's commencement.

Part II, dealing with construction contracts, contains sections covering both adjudication and payment. Section 108 deals with adjudication and this, together with four general sections and four supplementary sections, sets out the statutory provisions.

The first question is whether the adjudication provision contained within a contract complies with the Act. If one party relied upon the Scheme provision(s) rather than the express contract provisions, a delay would occur and the possibility of a speedy settlement would be lost amid complex arguments. This was addressed in the draft Scheme. Clause 19 stated that 'The first duty of the Scheme adjudicator is to establish to his own satisfaction that there is no effective adjudication agreement which satisfies the conditions in clause...'. In other words, unless there is some formal process of approving adjudication procedures that comply with the legislation, there is potential for the adjudication process to be disrupted. Although the revised Scheme contains different wording, the issue remains. It would seem that using adjudication procedures adopted by recognised contract authoring bodies is prudent rather than relying on bespoke adjudication provisions. However, there is no guarantee that such contracts would be better drafted in this respect. A rather bureaucratic way around this would have been for the Act to make provision for the approval of contract authoring bodies' adjudication provisions.

Construction contracts

The Act is only concerned with construction contracts and section 104 provides a definition. The Act defines a 'construction contract' as an agreement to carry out construction operations or to arrange for or provide labour for their carrying out. This also includes agreements to do architectural, design or surveying work and the provision of advice on building, engineering, decoration, or landscaping in relation to construction operations. This definition has a wide application and includes contracts for the appointment of consultants. Additionally, the definition may embrace collateral warranties where they relate to construction operations, although no one is entirely sure. However, the Act does not cover contracts of employment within the meaning of the Employment Rights Act 1996, but it would affect 'genuine' labour-only subcontractors. In cases where agreements cover both qualifying construction operations and other activities, the Act applies to the whole of the contract in all likelihood, although the point is not free from doubt.

A construction contract with a residential occupier is specifically excluded from the Act and its provisions are not applicable. Section 106(2) attempts to make clear what this means, by saying that it is a contract that 'principally relates to operations *on a dwelling* (authors' italics) which one of the parties to the contract occupies, or intends to occupy as his residence'. The words 'on a dwelling' suggest that one must pre-exist before the exclusion could operate. The term 'dwelling' in turn is also defined and means either a dwelling house or flat and consequently does not cover a building that is both. The wording has changed from that in the Bill and what started as an exclusion in respect of work to dwellings 'the whole or any part of which is subject of operations to which the contract relates' and which could have included the provision of the dwelling itself, has changed to one which on strict interpretation apparently excludes this possibility.

Section 117 deals with the application of the Act to the Crown. Part II applies to construction contracts entered into by or on behalf of the Crown, otherwise than in a private capacity.

In Section 104(7) it is stated that 'This Part applies whether or not the law of England and Wales or Scotland is otherwise the applicable law in relation to the contract.' This has two possible meanings. First, that it applies regardless which of these two legal systems is applicable to the contract, and secondly that it embraces all contracts related to projects in these countries, regardless of the governing law. The latter of these definitions is more obviously correct. Overseas companies building in the UK, even when using contracts based on different legal systems, would be caught by the Act.

Construction operations

Section 104 refers to construction operations (see section 105) and an extensive definition of what is and what is not a construction operation is provided. To be prescriptive in defining construction operations is dangerous, notwithstanding that this had been done under other legislation. The definition underwent some amendment during the Bill stage to take account of a number of matters, including representations by the process engineering industry for the exemption of plant and machinery and its associated steelwork.

Construction operation includes construction, alteration, repair, maintenance, extension, demolition or dismantling of buildings or structures forming or to form part of the land including any integral preparatory work. Also included is a specified range of civil engineering works, mechanical, electrical and other services, painting and decorating. The definition also covers cleaning to buildings and structures when carried out in connection with their construction, alteration, repair, extension or restoration. A list of exclusions is also provided.

The principal excluded activities are concerned with drilling and extraction work and plant and machinery work where the primary activity on site is processing engineering. The suppliers of construction materials and components did not wish their contracts to be covered by legislation and consequently the Act excludes contracts for the manufacture and delivery to site of building and engineering components or equipment, materials, plant or machinery, components for systems of heating, lighting, air-conditioning, ventilation, power supply, drainage, sanitation, water supply, or fire protection or for security or communications systems. However, where the contract also provides for their installation it falls within the ambit of the Act.

The issue of off-site fabrication is a bone of contention and some believe its exclusion from the legislation is an error because a significant proportion of work is now executed off site.

The production and/or installation of sculptures, murals and other works, which are wholly artistic in nature, are also excluded as construction operations. A further restriction included in the Bill but dropped from the Act was that of signwriting and erecting, installing and repairing signboards and advertisements. Presumably, they now fall outside the legislative framework.

The relative complexity of the definitions means that in practice there will be arguments.

Nature of agreement

Part 11 only applies to contracts in writing, a wide definition being adopted so as to ensure few contracts will be excluded. The definition of what constitutes 'in writing' is set out and an agreement in writing exists where there is an agreement whether signed or not or an exchange of written communications, or an agreement is evidenced in writing or by reference to terms which are in writing. This is a comprehensive definition but, in addition, the agreement may be recorded by any means or third party with the consent of the parties.

Where there is an exchange of written submissions in adjudication, arbitration or litigation, and it is alleged by one party that a contract other than in writing exists and this is not denied by the other party in their response, it constitutes an agreement to that effect.

Adjudication

Section 108 sets out a statutory right to refer a dispute to adjudication, 'a dispute' including any difference arising under the contract. The effect of this section is that a construction contract should contain an adjudication procedure that complies with sections 108(1) to (4). Where a contract does not comply with these requirements, the provisions of the Scheme for Construction Contracts will become operational as an implied term of the contract. Section 108 of the Act provides a legislative framework, which does not exclude the possibility of the incorporation of additional provisions to facilitate adjudication. Indeed, there is a need for such provisions which usually will be found in a set of adjudication rules. Such procedures are not to be inconsistent with the Act but simply to support its requirements.

Although under section 108(1) each party has a right to adjudication in respect of a dispute arising under a contract, are they bound to use the procedure to settle a dispute? Alternatively, can they agree subsequently to use some other method, including alternative dispute resolution. The whole purpose of the legislation is an attempt to provide an equitable and speedy resolution of disputes but if the parties choose not to implement the provisions and decide otherwise, it is a matter for them. However, any contract that provided for such a possibility before the dispute arose may fall foul of the Act by providing for something that is incompatible with the effect and/or spirit of the Act. If this occurred, the Scheme adjudication provisions would become operative.

This provision covers all disputes under the contract regardless of their nature, but there may be difficulties. Consider misrepresentation, rescission, negligence, conversion, and possibly common law damages. These are all excluded. Again, disputes over Value Added Tax or Statutory Tax Deductions under the Finance Act 1975 are probably excluded.

All the following requirements, that the Act states must be incorporated into a contract, are best drafted wherever practicable in the words of the Act. Subsection (2) sets out that a contract shall incorporate a range of provisions, as follows:

'(a) Enable a party to give notice at any time of his intention to refer a dispute to adjudication.'

It is clear that the right to adjudication can be exercised at any time and is not constrained by practical completion, making good of defects or indeed the final certificate. However, it must be intended that a notice can only be given once a dispute has arisen.

The form the notice takes can be agreed by the parties, failing which it can be given by any effective means (section 115(1)–(3). It is open to speculation what might constitute 'effective means' but section 115(4) does provide that notice by pre-paid post to the last known address, or where the addressee is a corporate body to the registered or principal office, is taken as being effective. Although the form and means of delivery of the notice are included in the Act, they are not specifically required to be incorporated in the contract. However, as the parties may agree the form the notice takes they may well include specific provision covering these matters, possibly including the use of a standard form, and specifically the date when the notice is taken as served.

It is possible for a dispute to arise between the parties and for a notice to be given without the knowledge of one of the consultants on the project. As the dispute could result from their action, it would seem good practice to provide them with a copy of the notice. Although, this does not offend the Act, the adjudication cannot be delayed because of the consultant reviewing any relevant decision, as this would be inconsistent with the Act.

'(b) Provide a timetable with the object of securing the appointment of the adjudicator and referral of the dispute to the adjudicator within seven days of such notice.'

This provision probably envisages a separate action to that of giving notice of intention to refer the dispute. This raises two separate but important points. First, would the pre-appointment of an adjudicator comply with the Act, and secondly, can this provision be fulfilled simultaneously with giving the notice of intention to refer a dispute. The question of whether section 108(2)(b) requires the appointment of an adjudicator subsequent to a dispute meets the practical point of it being better to have the appropriate horse for the course than to have a pre-selected adjudicator.

It would seem there is at least an argument that an arrangement made prior to the dispute may not accord with the Act. If the operation of section 108(2)(b) is sequential to section 108 (2)(a), a pre-appointment may be invalid, but there is nothing which expressly prevents a timetable from being stated in the contract and which provides for a named adjudicator, so long as the timetable provides for the dispute to be referred within seven days of the notice given under section 108(2)(a). The position adopted by the Scheme which enables the parties to name an adjudicator in the contract or subsequently would seem to confirm this view.

If a named or pre-appointed adjudicator is not used the Act does require that a speedy process of appointment needs to be followed. The period for the commencement of adjudication was added to later drafts and is an essential requirement if the quick resolution to disputes is to be fulfilled. However, failure to enter into an adjudicator's agreement within the specified time need not be calamitous, provided the contract requires such an appointment within seven days, as required by the Act. In giving the notice of intention to refer a dispute, especially where the adjudicator is not named or pre-appointed, it would be desirable to provide a summary of the dispute together with the documents and other information to be relied upon. This would facilitate the correct appointment of an adjudicator, as would the use of a standard form of adjudication agreement. Neither of these matters are covered by the Act.

Another question this poses is whether an adjudicator can be appointed before an adjudication agreement is concluded. If there is the presumption that they are separate and hence the appointment can be fulfilled prior to entering into an adjudication agreement, there is no problem. There is, however, a good case for arguing that an appointment is not made until the adjudication agreement is entered into because until that time there is no adjudicator available to resolve the dispute. If this is so and no other provisions exist within the contract, it would be possible for a party to impede the formation of an adjudication agreement and this would defeat a main purpose of the Act. A different situation arises where the adjudicator is nominated in a way similar to that of an arbitrator. Regarding the second point, it would appear that there is nothing that precludes the notice of intention and the referral of the dispute being given simultaneously.

'(c) Require the adjudicator to provide a decision within twenty-eight days of the referral or such longer period as is agreed by the parties after the dispute has been referred.'

The 28-day period during which the adjudicator must give a decision is determined by the date on which the notice of referral is served, and in accordance with section 116(1) begins immediately after that date. Section 116(3) identifies the usual public and other holidays that are excluded for this purpose but does not exclude weekends. The date of the notice will usually be the date on which the notice is received but as the parties can determine what constitutes service, it does not follow that this will invariably be the case. In the absence in the contract of what constitutes service, the Act provides that service is effective when the notice is delivered, but again this need not accord with the actual receipt by the party.

This 28-day period may be extended by agreement of the parties but only after the dispute has been referred. The agreed extension can be of any length. Nevertheless, there is no question of making an agreement before the dispute is referred as this would be invalid, no doubt because the essence of speed could be immediately lost. The overall anticipated maximum time-scale under the Act for providing an adjudicator's decision is 35 days and some observers believe this is too long and does not fulfil the level of speed wanted by many disputants. However, even this period can be appreciably extended because of a failure in the appointment of an adjudicator.

It would not be appropriate to permit the adjudicator not to make a decision. Where an adjudicator fails to provide a decision or declares his decision is that he cannot decide, the adjudication procedure could be recommenced by appointing a new adjudicator.

In addition to the requirement to reach a decision within the stipulated 28 days, it would be sensible to provide that the adjudicator then communicates this decision immediately to the parties, subject to any lien on the decision to secure fees. This may cause a delay beyond the 28 days but there is no requirement under subsection 2(c) that the decision must be both reached and communicated. Consequently, it would seem that an adjudicator could, unless otherwise stated in the adjudication agreement, require the adjudication fee to be paid before releasing the decision reached. This makes commercial sense.

'(d) Allow the adjudicator to extend the period of twenty-eight days by up to fourteen days with the consent of the party by whom the dispute was referred.'

The adjudicator may require longer in which to provide a decision and therefore should be allowed to extend this period if the consent of the referring party can be obtained. Interestingly, some observers take the view that the adjudicator should not need to seek approval of the party bringing the dispute. Nevertheless, this provision is not surprising, in that it will have the effect of keeping to the prescribed period unless the party who feels aggrieved is content to extend this time. No reason need be given if the party does not wish to agree to extend this period, but before declining to do so it would be prudent to consider the possible effects. Presumably an adjudicator would not seek to extend time without good reason. In any event, an adjudicator's decision is unlikely to be considered invalid just because it was reached outside the stipulated period. Bearing in mind that the claimant can determine his own time in which to prepare a claim prior to notification, it would seem that the responding party could be put at some disadvantage in complying with these timescales.

'(e) Impose a duty on the adjudicator to act impartially.'

The idea that an adjudicator should or could act in any other way than impartially is surely implausible. There is no requirement that the adjudicator is independent but this is practically essential, otherwise the contract supervisor could be appointed. Where a potential conflict of interest exists, even the declaration of this may be inadequate to allay fear of possible bias and is likely to produce a subsequent challenge to the decision. The perception of an adjudicator's impartiality and independence, in the eyes of the parties, is as important as the fact itself.

'(f) Enable the adjudicator to take the initiative in ascertaining the facts and the law.'

The adjudicator can act in an investigative manner and not be restricted to the submissions of the parties. The words mirror the Arbitration Act 1996. The adjudicator is given a wide discretion in seeking to establish the facts and the law. There is no reference to the adjudicator using his own technical experience, although in practice this will often occur. In such circumstances he should inform the parties and invite further comment. The adjudicator can set the procedures to be followed. Institutional rules may give examples of what the adjudicator is able to do.

Any contract clause providing for an adjudication must be clear and not be able to be construed as an arbitration clause. It must not be final and binding. Therefore, it may be desirable in any contract to state that the adjudicator is acting under section 108 of the Act and not as an arbitrator, in order to avoid any possible confusion, although the courts would construe the meaning of the clause.

Section 108(3) provides that the adjudicator's decision is final and binding until the dispute is finally determined by litigation, by arbitration where such agreement exists or is otherwise agreed, unless the parties agree to accept that the adjudicator's decision as final.

The Construction Industry Council was opposed to the Bill making the adjudicator's decision final and binding and achieved an amendment which removed this from the legislation. Again, this removed the spectre of statutory adjudication being arbitration. The 'final and binding' nature was considered to be arbitration and this view was reinforced by the fact that the draft 'Scheme' (but not the approved version) referred to an adjudicator's 'award' and not 'decision', thus implying (probably correctly) that this form of adjudication was arbitration by another name. Although this particular spectre has now gone, there are some industry members who are unhappy that there are still references to the Arbitration Act in section 108(6) of Part II of the HGCRA 1996 and the merging of concepts.

Under the Act, the adjudicator's decision has temporary effect until the dispute is finally determined by legal proceedings, arbitration or by agreement of the parties. This last point enables the parties to set the adjudicator's decision to one side where they so agree.

There is no restriction under the Act as to when the adjudicator's decision can be challenged, and subject to any restrictions under the construction contract, this can be either before or after practical completion; indeed, before adjudication commences the adjudicator's jurisdiction can be challenged.

There is no express requirement in the Act that the adjudicator's decision should be implemented or how to achieve it, but it is necessary that this is set out in detail in the contract. Notwithstanding that the decision is binding until successfully challenged, it is also essential to know the courses of action available if there is a failure to implement the adjudicator's decision.

Adjudicator's immunity

Section 108(4) requires the contract to provide that the adjudicator shall not be liable for anything done or omitted in the discharge of functions unless due to bad faith. It also requires that any employee or agent of the adjudicator is similarly protected from liability. An adjudicator's potential for liability is substantially reduced as a consequence of this provision as they have a measure of protection against wrong decisions, regardless of negligence. However, this immunity is provided as part of the contractual machinery, notwithstanding the reference in the Act, and does not extend to providing statutory immunity. Statutory immunity is something many observers wished to see, but the main argument, that without some immunity it may be difficult to recruit adjudicators, was accepted by the Government.

Reference to Parliament and the Scheme for Construction Contracts

The Bill as proposed to Parliament gave certain powers to the Secretary of State which enabled orders to be given to add, amend or repeal what constituted 'construction operations' and to amend the definition of 'dwelling'. However, no order could be made until approved by each House of Parliament. This position was maintained in the Act but also extended to cover the addition, amendment or repeal of what agreements constitute construction contracts. Exclusion orders have been made subsequently and are discussed in the next section.

A further power provided by the Bill was for the minister to introduce delegated legislation in the form of a scheme (the Scheme for Construction Works), with the requirement to consult 'such persons as he thinks fit'. The Act accordingly provides that such regulations should have to be approved by each House of Parliament. This provides industry with a measure of protection from having totally repugnant requirements imposed. No express provision is contained covering additions, amendments or repeal of the Scheme but the ability to use regulations to make the Scheme will no doubt be deemed properly to include these, subject of course to consultation and approval of both Houses of Parliament.

Many parts of the industry and elsewhere focused on the initial draft Scheme and this concern prompted a complete review of its provisions, notwithstanding that it was and still is possible to avoid its application by making adequate provision for adjudication in the contract.

Exclusion orders

On 6 March 1998, statutory instrument No. 648/1998 was made and under article 1 (1) became operative eight weeks later, on 1 May 1998. This order was given under sections 106(1)(b) and 146(1) of the HGRCA and applies to England and Wales, with a similar order being made in respect of Scotland. The effect of this order is to exclude certain types of contracts from the operation of Part II of the HGRCA, and the Act identifies four types of contractual agreement:

- agreements under statute
- private finance initiative (PFI)
- finance agreements
- development agreements.

This covers a fairly wide range of contracts and with the increasing use of the PFI will exclude much construction activity in the future. Surprisingly, millennium sponsored projects with their particular problems are not excluded.

Agreements under statute

This only covers a very small number of statutes and these are stated in article 3 to be those under sections 38 and 278 of the Highways Act 1980, sections 106, 106A and 299A of the Town and Country Planning Act 1990, and section 104 of the Water Industry Act 1991. Additionally, externally financed development agreements under section 1 of the National Health Service (Private Finance) Act 1997 are excluded.

Private finance initiative

Construction contracts entered into under the private finance initiative as set out in article 4 are excluded. For a contract to be considered PFI it must comply with a number of principles set out in articles 4(2)(a)(b). Additionally, one of the parties to the contract must be the government, a government department or an authority or organisation as set out in article 4(2)(c).

Finance agreements

Agreements which are primarily concerned with financing the works are excluded. Such agreements embrace contracts of insurance and other contracts whose principal purpose is the setting up or dissolving of a company, association or partnership. The transferring of securities, the lending of money or acting as surety are all excluded.

Development agreements

Construction agreements which are development agreements within the order are also excluded. These contracts involve a grant or transfer of land on which the principal construction operations are to take place and to which the contract relates. This covers the transfer of the freehold and also the grant of a leasehold where the period is for at least one year beyond the completion of the construction operations included in the contract. This means that turnkey contracts and some design and build contracts will be excluded from the provisions of the HGRCA where construction operations are set out in an agreement for lease or development agreement.

Approved Scheme for Construction Contracts

The DOE published a consultation paper, *Making the Scheme for Construction Contracts*, during the autumn of 1996 in order to consult industry as to the content of the Scheme for use in England and Wales, with a view to the Act taking effect early in 1997. However, delays occurred and the pre-consultation with the construction industry umbrella groups and the subsequent general consultation periods were extended. A final draft of the Scheme was then produced by the DoE without further consultation with the industry but subject to legislative approval as provided in the Act. The Scheme for Construction Contracts Regulations 1998 was made on 6 March

1998 and came into force on 1 May 1998, thus triggering Part II of the HGRCA.

Application of the Scheme

Where a construction contract does not comply with the requirements of the HGRCA the Scheme becomes operative. However, the way the Scheme applies is dependent on whether there is non-compliance with section 108 dealing with adjudication or sections 109–111 and 113 dealing with payment. Any non-compliance with the adjudication requirements of section 108 means that the whole of Part I of the Scheme will apply regardless. A small failing in the contract provisions concerning adjudication means the whole adjudication provision is invalid and the Scheme provisions apply. By contrast, non-compliance with the payment provisions means the relevant part only of Part II of the Scheme will apply. Where the contract contains ineffective payment provisions the Scheme simply fills in the gaps or replaces the offending provision.

The Scheme is not well drafted, being complex and verbose, and is best avoided if possible. For example, it does not address matters such as what happens if the time periods are not complied with, there is no provision for the correction of a decision and the responding party does not have an express right to make a response.

Part One – adjudication

The Scheme sets out 26 paragraphs that apply if the construction contract fails to comply with section 108 of the HGRCA. Part One covers four main areas that include the notice to seek adjudication, the powers of the adjudicator, the adjudicator's decision and the effect of the decision.

Notice of intention to seek adjudication

The Scheme confirms the right of any party to a construction contract to refer a matter to adjudication and requires that notice is given in writing to all other parties to the contract. Parties to a contract are those bound by the contract. This would not include the supervising officer, even if it were desirable to provide a copy of the notice to the person performing this role. This written notice is known as the 'notice of adjudication' and should set out the nature and details of the dispute and the redress sought, together with the names and addresses of the parties to the dispute.

Once the notice of adjudication has been issued the party referring the dispute shall commence the procedure for putting the adjudicator in place. This can be by agreement of the parties but will often be by giving notice to the person specified in the contract. Where no person is named or they are unable to act, the contract provisions for getting the nominating body to make an appointment will be implemented. If the contract does not contain such provisions, the referring party shall request an adjudicator nominating body to make a selection. There is no approved list of adjudicator nominating bodies, although that was the original intention. An adjudicator nominating body can be any organisation that holds itself out to the public for making such appointments, so long as it is not a party to the dispute (see Appendix 3 for a list of nominating bodies). This provides a freedom of choice and also prevents the situation whereby an approved list requires updating to avoid ossification.

Where an adjudicator nominating body is required to select an adjudicator, its decision must be communicated to the referring party within five days of receiving the request. If this is not done, the referring party may request another nominating body to make a selection or, alternatively, agree with the other party who shall act. The adjudicator, once requested to act, must indicate within two days whether or not the appointment will be accepted. Any failure to do so, or where a previously named adjudicator indicates an unwillingness to act or is unable to act, the referring party may recommence the procedure in accordance with paragraph 6(1). There is a choice: either request the nominating body specified, use any other nominating body or request another person specified in the contract to act as adjudicator. Any of these options could be used, if available.

The person selected as adjudicator is carefully controlled. Paragraph 4 requires that the adjudicator must be a person acting in his personal capacity and not an employee of any of the parties to the dispute. A company cannot therefore be appointed as adjudicator but someone with a conflict of interest may be, as long as this is disclosed. How the declaration should be made is not covered, but it would seem prudent for an adjudicator to inform all parties, otherwise there is the increased danger of a subsequent challenge to any decision given. Where a party objects to the appointment of a particular person as adjudicator it will not invalidate the appointment nor any decision made by the adjudicator (paragraph 10).

Under paragraph 7(1), the referring party shall within seven days of the notice of adjudication refer the dispute to the adjudicator if selected in accordance with paragraphs 2, 5 or 6. This provision refers to selection and not acceptance by the adjudicator, therefore the referral of the dispute will frequently take place before it is known whether the adjudicator will act. Moreover, the dispute only has to be referred where a selection has been made and this, for a variety of reasons, can be much later than seven days from the notice of adjudication. When the referral notice is sent it shall include copies of all relevant documentation that are to be relied on, and a copy of the notice and documentation shall be sent to the other parties.

Parties may agree that an adjudicator can act on multiple disputes under the contract or on related disputes under separate contracts. They may also agree to extend the period for making a decision.

Termination of adjudicator

The adjudicator may resign at any time and must resign where the dispute is the same or substantially the same as a dispute already adjudicated. This latter provision avoids the situation whereby a party can keep referring disputes in the hope of achieving a different decision. An adjudicator should also cease to act where he becomes incompetent because the dispute varies significantly from that referred.

In circumstances where an adjudicator chooses to resign, the referring party may serve a further notice of intention to refer the dispute and shall then request the appointment of a new adjudicator in accordance with the procedure discussed above. Where a new adjudicator is appointed, he may request copies of all information provided at the earlier adjudication and the parties are required to provide such information insofar as is practicable. The parties may agree to terminate the appointment of an adjudicator at any time.

Powers of adjudicator

The adjudicator can take the initiative in determining disputes and may adopt any appropriate procedure consistent with the terms of contract and applicable law. This suggests that the adjudicator must act in accordance with the law and cannot apply a decision otherwise than in accordance with the law (he cannot act as amiable compositeur). In determining the procedures, the adjudicator must avoid unnecessary costs; the presumption must be that if unnecessary costs are incurred they are irrecoverable from the parties. It would also appear that if the parties were of the view that unnecessary costs would be incurred, they should tell the adjudicator and seek an alternative approach.

Although the adjudicator can act in any appropriate manner, paragraph 13 sets out a list of specific powers. This list includes a range of obvious matters such as:

- requiring further documents and written statements supporting or supplementing other documents,
- obtaining representations as considered necessary,
- subject to notifying the parties, appoint experts and the like,
- make site visits and carry out tests,
- give directions as to time limits.

The ability to make site visits and inspections and to carry out tests and experiments is said to be subject to third party consent(s).

Other powers include the ability to:

- decide the language to be used in the adjudication and whether translation of any documents should be made,

- meet and question any of the parties to the contract and their representatives,
- take into account any other matters which the parties agree, notwithstanding that they have not been included in the referral notice,
- take account of any matters under the contract which are considered 'necessarily connected with the dispute,
- open, review and revise any decision taken or certificate given under the contract, except where it is stated to be final and conclusive.

The issue of language is not likely to be encountered frequently but it may be from time to time, as the HGRCA applies, subject to exceptions, to all construction works in the UK, regardless of whether one or both parties are from overseas.

The power of adjudicators to meet any of the parties or their representatives is interesting in so much as one party alone or each party separately can be questioned. This procedure is common in ADR but not in processes which are 'quasi-judicial'. Whether it is good practice under an adjudication in accordance with the Scheme will be debated, but if it is used the information released should be 'open tabled'. The opportunity for parties to be assisted by advisers is maintained unless it is excluded by an agreement by the parties. There are conflicting views regarding the use of advisers in adjudication. They may increase the cost and complexity of the dispute. Too many advisers are a bad thing. A good adjudicator has the powers to control the dispute resolution process and additionally can restrict representation to one person (section 16(2)), thus avoiding the problems of increased cost and complexity.

The ability to open and review decisions only applies to those made by the persons administering the contract and does not extend to include any made by an adjudicator (paragraph 9(2)).

Parties must comply with the adjudicator's directions and other requirements and where without reasonable cause they fail to do so, the adjudicator may still proceed with the adjudication. It is never likely to be in the best interests of a party not to comply because the adjudicator is permitted to draw whatever inferences are justified in his opinion.

Paragraph 18 requires the adjudicator not to disclose any information or document supplied during the adjudication except for the purposes of the adjudication itself. However, parties should be aware that this only applies to documents and the like that the party has indicated are to be treated as confidential. If parties do not wish the information to be subsequently disclosed they should mark each document 'confidential and for the purposes of the adjudication only' or some similar words.

Adjudicators, together with any of their employees or agents, are free of liability for anything they do, or do not do, in the discharge of their functions in the adjudication, except where there is bad faith. As this immunity is provided under delegated legislation it will extend to third parties and in this respect may well differ from some other adjudications not conducted under the Scheme.

Adjudicator's decision

As discussed earlier, the adjudicator must reach a decision that is in accordance with the applicable law in relation to the contract. Even where a party fails to comply with the adjudicator's requirements in respect of providing information, the adjudicator can still make a decision based on the information provided. In circumstances where the information is provided late, the adjudicator can attach whatever weight he thinks appropriate.

The decision of the adjudicator must be given within the period set out in paragraph 19. This is usually 28 days after the date of the referral notice but this may be extended to 42 days if the referring party agrees. Both parties can agree to extend the 28-day period by any length but can do so only after giving notice of the dispute. If the adjudicator fails to meet the deadline for the decision any party may serve a fresh notice of dispute and the whole process is recommenced. In these circumstances, the new adjudicator may request copies of all information provided at the earlier adjudication and the parties are required to provide such information insofar as it is practicable. Here the new adjudicator can act in the same way as where the previous adjudicator had resigned.

The adjudicator's decision shall cover all matters of dispute referred to in the notice of referral and may issue separate decisions on each aspect of the dispute. Decisions may be given at different times but each of them must comply with the periods referred to or agreed under paragraph 19(1). The decision itself can be a review of any decision or certificate of any administrator under the contract or a determination of who is liable to make a payment and whether any interest is payable and at what rates. Decisions relating to payment may also specify when the payment is due and the final date for payment, subject to the requirements of section 111(4). Either party may also request that the adjudicator provides reasons for the decision and where such a request is made the adjudicator is obliged to provide them. The adjudicator is required to give a decision and this must address the matter referred; the adjudicator cannot frame a decision in terms such as 'further information' is required. Under some contracts in the past an adjudicator could decide that payment be made to a stakeholder but this is no longer an option because it does not provide a decision as such. Such an option would be, if drafted, outside the Act.

Once the adjudicator has reached a decision it shall be notified to the parties and they shall comply with it immediately it is received, unless the adjudicator has directed otherwise. This decision is final and binding and the parties must comply with it until the matter is finally determined by legal proceedings or arbitration, if provided for in the contract or agreement.

Paragraph 24 provides that section 42 of the Arbitration Act 1996, suitably modified, is applicable to the Scheme for Construction Contracts. This means that the court is given powers to make an order to force compliance with peremptory orders made by the adjudicator. The Scheme amends the section by deleting subparagraph (c) of subsections (2) and (3) and this ensures that the power of the court cannot be excluded by the parties agreeing otherwise. Furthermore, the second amendment made means the court does not have to satisfy itself that the applicant has exhausted the adjudication process before the court will act. Although the adjudicator has wide powers that can be exerted on a recalcitrant party, it may be that the remedy required is inappropriate for the problem and that the court's competence to imprison may be required.

Payment of adjudicators

Under paragraph 25, the adjudicator is entitled to payment of reasonable fees and expenses incurred in the adjudication and the parties are jointly and severally liable for any such amounts. The Scheme allows the adjudicator to determine his own fees but prudence dictates that at least the hourly rate is determined before the adjudicator is appointed. Obviously this may cause difficulties because of the short time allowed for making the appointment. Where both parties are unhappy with the fee the problem is lessened, but if only one party is unhappy the adjudicator really is in a very strong position. This lack of a fee structure should be addressed either at the stage of naming the adjudicator or by ensuring that the nominating body has one published.

Fees and expenses are still recoverable where the adjudicator is required to resign because the dispute has already been determined or where the adjudicator cannot act because the dispute varies significantly from that referred. The same position prevails where the parties themselves terminate the adjudicator's appointment, except where the termination is brought about by the adjudicator's own default or misconduct. In the latter circumstances the adjudicator cannot recover either fees or any expenses incurred.

Other matters related to the statutory need for adjudication provisions

Claims under contracts can lead to payments being made because someone is considered at fault and this may extend to consultants working on the project, not least because their own contract with the client will generally be subject to the HGCRA. The statutory provisions concerning adjudication will then apply and if implemented they may lead to a loss being incurred. Those who suffer such loss may seek protection from their professional indemnity insurance, but will they be covered where the 'damage' is determined in an adjudication? This matter has been given much consideration and the situation will take some time to settle down because there is uncertainty over several points. First, does an adjudicator's decision given under the HGCRA create a liability? After all, the decision is only 'temporary', in that the dispute may be referred subsequently to arbitration or litigation depending on the contract. Nevertheless, a loss may be suffered in the interim, even though a decision is subsequently altered. Secondly, is an adjudicator able to reach a commercial or technical decision rather than one based on legal grounds? What will be the attitude of the various insurers if a commercial decision is reached? Because of the shortness of time to arrive at a decision, adjudication is thought by many to offer only a rough-and-ready solution without the added dimension that it may circumvent the law. If the Scheme for Construction Contracts applies, then the position is clear in that the adjudicator is required to reach a decision 'in accordance with the applicable law in relation to the contract', but other rules of adjudication may provide otherwise. It is doubtful whether adjudication can be extra judicial. If the parties agree that it be so, it ceases to be adjudication.

Insurers may write policies that exclude cover for loss arising from adjudication or specific types of adjudication, and consequently there is need for diligence both in performing one's work and arranging insurance. Alternatively, insurers may require that the adjudicator's decision must be 'appealed' before they are required to indemnify the party insured under the policy. The type of policy and its wording can be crucial and everyone who may be involved with adjudication, in whatever way, should ensure the policy provides appropriate cover. Many policies require notice to be given of events that may lead to a claim and in this regard it would be prudent to give notice to the insurer even before a dispute is referred because of the possibility of counterclaims emerging.

Chapter 10
JCT Standard Forms of Building Contract and Associated Subcontract Forms

Generally

This chapter begins with a brief discussion on the historical developments of the JCT 1980 Standard Form of Contract and the JCT 1981 With Contractor's Design Form of Contract, regarding dispute resolution. A commentary follows on the JCT amendments issued in 1998 so that the contracts would comply with the provisions of the HGRCA. The latter part of the chapter then deals with subcontracts associated with JCT 80, including those not published by the JCT. Additionally, it briefly covers the 'Green' and 'Blue' subcontract forms because of the historical significance of the developments reflected in these contracts.

JCT Standard Form of Building Contract 1980 (JCT 80)

In 1980 the Joint Contracts Tribunal published a significantly revised Standard Form of Building Contract together with variants of this form. However, despite these major changes the contract made no provision for the settlement of disputes other than by arbitration. As with its predecessor form, JCT 63, there was no provision for alternative dispute resolution or indeed for adjudication. ADR was not seriously mentioned in England until nearly a decade later and is still not part of JCT Amendment 18 (JCT 80) or Amendment 12 (JCT 81 and IFC 84).

At the time of publication of the earlier editions, disputes under building contracts, particularly between client and contractor, were less frequent than they are today. When they occurred they were generally settled without recourse to litigation, arbitration or other formal means, no doubt facilitated by the generally accepted patriarchal role of the architect. Historically, when such a dispute arose, the architect would act in a conciliatory and adjudicative manner. Where this failed, it was appropriate to move directly to a formal forum and consequently there was only provision for arbitration.

The settlement of disputes in article 5 of the contract is not substantially different from clause 35 of the 1963 Edition. This article restricted the commencement of arbitration, except in the case of certain prescribed events, until after completion of the works. The idea that not all problems required early resolution still prevailed. There was a distinction between those problems requiring early resolution and those that could wait. The distinction arises from the questionable belief that unless certain issues were resolved, contract completion was unachievable. Surely any dispute should be resolved immediately to avoid the development of an unhelpful working environment. Curiously, the JCT never adopted a procedure for the architect to settle disputes in a way similar to that contained in the ICE Forms of Contract[1].

In July 1987, the issue of amendment 4 to the JCT 80 included many changes to the provisions dealing with the settlement of disputes. Article 5 was substantially reduced and a new clause 41 incorporated. Clause 41 provided for the appointment of the arbitrator, and where parties were unable to agree on the appointment this was to be made by the president or vice-president of the RIBA, RICS or CIArb as identified in the appendix. This change provided a wider choice of appointing organisations and was perhaps more to do with the status of the organisations than with providing a real choice for the user. Nevertheless, the very existence of this choice will no doubt have an influence on the nature of arbitration, simply by the involvement of arbitrators from a more diverse range of professional backgrounds. If no selection is made in the appendix and the parties cannot agree on the arbitrator, the president or a vice-president of the RIBA becomes the appointor.

The award of the arbitrator was final and binding on both parties but subject to clause 41.6, permitting either party to appeal to the High Court to determine questions of law which arose during the arbitration or which arose out of the award (sections 1(3)(a) and 2(1)(b) of the Arbitration Act 1979).

July 1987 saw a major review of the provisions dealing with the settlement of disputes. Consequently, the issue in July 1988 of amendment 6 to the JCT 80 amended parts of the arbitration clause. An amendment to clause 41.1 required that either party could give a written notice to the other party when they required a dispute to be referred to arbitration. Before this amendment, clause 41.1 referred to a 'written request to concur in the appointment of an arbitrator' but did not require that a notice should be given to commence the arbitration; therefore the timing of the actual reference to arbitration was uncertain. In addition, a new provision in clause 41.2.1 required that the arbitration should be conducted under the JCT Arbitration Rules, which were first published on 18 July 1988. Understandably, JCT 80 did not comply with the requirements of section 108 of the HGCRA regarding adjudication and the Joint Contracts Tribunal produced revised provisions (discussed later in this chapter), in order to avoid the application of the 'Scheme for Construction Contracts'.

JCT Standard Form of Building Contract With Contractor's Design 1981 (JCT 81)

This contract form followed much the same development as JCT 80 in respect of dispute resolution, until the introduction of the 'supplementary provisions' in 1988. Until 1988, JCT 81 only provided for disputes to be referred to arbitration and these provisions, except for the exclusion of the joinder provisions, were much the same as JCT 80.

In February 1988, amendment 3 was issued together with guidance notes. This amendment incorporated optional supplementary provisions as the result of external pressures from, among others, the British Property Federation (BPF). Supplementary provisions S1 to S7 were published but S1 dealing with adjudication is the only one that is relevant to this book. The introduction of clause S1, albeit optional, broke new ground for the Joint Contracts Tribunal in that this main contract form specifically allowed for adjudication to be adopted as a means of dispute resolution. Oddly, the approach and use of an adjudicator in this way did not extend to other JCT standard forms of contract.

In order for the supplementary provisions to apply, this must be stated in the appendix and deleting the words 'not to apply' does this. If the provisions are not to apply, delete the words 'to apply'. An act by the parties is required for either condition and if this is not done because of an oversight an unclear position emerges. In these circumstances, it would seem that the supplementary provisions would not apply and any subsequent attempt to impose them could be resisted. However, there is nothing to prevent the parties from adopting the adjudication provision later, following an oversight, if they so agree.

The purpose of the adjudication provision is to enable speedy settlement of disputes by an expert, as they arise during the course of the works. The BPF and others felt this provision to be beneficial. No longer was there a need to wait until practical completion before referring many disputes to arbitration. Furthermore, it also attempts to avoid the formal proceedings of arbitration and to introduce a simpler and cheaper procedure. There is nothing to prevent arbitration prior to practical completion where the parties agreed. Perhaps the most interesting aspect of the provisions is that the adjudicator is an expert and not an arbitrator, thus enabling the introduction of the expert's own knowledge and experience. This would allow for a practical resolution of the dispute, uninhibited by the formal constraints relating to the law of evidence. Such a decision would not have the force of an arbitrator's award and be subject to the same enforcement procedures.

Where S1 of the supplementary provisions is applicable article 5, which deals with the settlement of disputes, is modified in its operation. Clause S2 sets out a list of 'adjudication matters' and provides that where any of these events occurs prior to practical completion or alleged practical completion, termination or alleged termination or abandonment of the works, then it shall be referred to the adjudicator.

If such a dispute or difference arises, either party may give notice to the other that they wish the adjudicator to decide the matter in dispute. There is then a 14-day period during which the adjudicator must be given the statements setting out the matter(s) on which a decision(s) is required. Within 14 days (or as otherwise agreed) of receiving these statements, the adjudicator is required to give a date when the decision can be expected. The adjudicator may require further information in order to reach this decision. There is no prescriptive timescale within the contract for the provision of this information or for the delivery of the decision, only a requirement to provide an expected date and the period for doing this can be extended where the parties agree. If a party fails to provide any information requested, the adjudicator, notwithstanding its absence, may give a valid decision.

A decision of the adjudicator is deemed to be a provision of the contract and is referred to as an 'adjudicated provision' which shall be final and binding on the parties unless referred to arbitration. Where the adjudicated provision conflicts with other contract provisions the adjudicated provision will prevail. In circumstances where one party is affected adversely, reference to arbitration will inevitably occur. Therefore, the adjudicator's decision has the force of contract and if the decision is not acceptable, either party may within 14 days of its receipt refer the matter to arbitration. Nevertheless, the adjudicator's decision remains effective as a contract provision. Arbitration will normally occur after practical completion but the appointment of the arbitrator will be made immediately to avoid any undue delay. Once the arbitrator has been appointed either party may request that orders and directions be given in a further attempt to facilitate a hearing as soon as possible on completion. If the parties require, the hearing itself could be prior to practical completion. In any subsequent arbitration hearing, the adjudicator may also be called to give evidence.

The adjudicated provision itself may also be the subject of a dispute or difference and where this occurs the supplementary provisions and arbitration provisions are still available.

The adjudicator is named in the appendix, and clause S1.5 provides for what happens where this is not done or where the named adjudicator is unwilling or unable to act. The adjudicator's fee is shared equally and may be required to be paid before the decision is given. All other costs related to the adjudication are borne by the party incurring them.

The introduction of the supplementary conditions to the JCT Standard Form of Building Contract With Contractor's Design 1981 was a landmark in that it was the first JCT contract to provide for an adjudicator across the whole spectrum of disputes. These provisions are broadly compatible with the spirit of the HGCRA but amendments have been published by the JCT.

Amendment 18 JCT Standard Form of Building Contract 1980

Prior to the HGCRA and after its royal assent the Joint Contracts Tribunal considered their own contracts and what was required so that they would comply with the requirements of the Act. These deliberations took place over an extended period to achieve consensus and because the JCT wished to ensure that their provisions stood up and the 'Scheme for Construction Contracts' would not need to apply. In November 1996 provisions were presented to the Tribunal, yet the final provisions were not published until

April 1998. Amendment 18 to JCT 80 includes amendments arising out of the proposals contained within *Constructing the Team* and to comply with the HGRCA. The amendment is lengthy and in parts confusing and has been criticised for its complexity and prolixity. The following discussion is concerned only with the amendments made in response to the requirement for adjudication under the HGRCA.

Articles of agreement

The articles of agreement have been amended so that any dispute or difference that arises under the contract may be referred to adjudication by either party. The adjudication is to be in accordance with clause 41A, which sets out totally new provisions.

In addition, two new articles allow disputes to be referred to either arbitration (article 7A) or legal proceedings (article 7B). The pros and cons of arbitration and litigation are dealt with in Chapter 2, and guidance notes to amendment 18 address the same issue. Guidance note paragraph 4.6 explains that the clause was drafted to allow the court to decide a dispute and to avoid the problems posed by *Northern Regional Health Authority* v. *Derek Crouch Construction Company Limited* (1984). The House of Lords judgment in *Beaufort Developments Ltd* v. *Gilbert-Ash Ltd and Others* (1998) overruled *Crouch* and leaves the JCT amendment appearing now out of date. Under amendment 18, which of the two procedures applies is determined by what is included in the appendix to the conditions. If the reference to 'clause 41B applies' has not been deleted then arbitration under article 7A is applicable. Where arbitration is not required, the words 'clause 41B applies' must be deleted. Both litigation and arbitration are subject to prior adjudication taking precedence if required by either party. Parties are not precluded by the Act from choosing not to exercise their right to adjudication but a right must be available. Adjudication is available for all disputes 'under' the contract but this does not extend to those 'in connection therewith'. Consequently some disputes, including those as to whether a contract exists, negligent misstatement, rectification, misrepresentation and common law damages claims, may be outside an adjudicator's power and referred directly to arbitration or the courts.

Article 7A provides for arbitration if the dispute is not settled by adjudication whether arising during or after the works are completed, abandoned or determined. Arbitration covers all disputes except those concerned with statutory tax deductions and value added tax. More importantly, it also excludes disputes about an adjudicator's decision, attempting to ring fence such a decision from instant attack. Summary enforcement in the courts of an adjudicator's decisions is available; enforcement cannot be the subject of arbitration under these provisions. Any arbitration under clause 41B shall be in accordance with the JCT 1998 edition of the Construction Industry Model Arbitration Rules (CIMAR).

Article 7B provides that any dispute may be the subject of legal proceedings and clause 41C will then apply.

Definitions and interpretation

The amendment also includes a number of additional definitions and these include the terms, adjudication agreement, adjudicator, party, parties, and public holiday. Clause 1.5, which deals with the giving of notices and is new, only refers to notices that are not elsewhere described and consequently does not apply to adjudication, which deals with this separately in clause 41A.4.2. Clause 1.6 is also new and deals with how the number of days is reckoned and partially accords with section 116 of the HGRCA.

Part 4: Settlement of disputes – adjudication – arbitration – legal proceedings

A totally new set of provisions is included in the amendment and this is mirrored in the new heading of Part 4.

Clause 41A covers the adjudication provisions and where a party exercises their right, this process will apply. These provisions together with the adjudication agreement provide the framework for adjudication and no separate rules are published or need to be incorporated.

Appointment of the adjudicator

The adjudicator mentioned in article 7A will decide any disputes that are referred to him. The JCT does not favour naming an adjudicator in the contract because the person appointed at commencement of the works may not be appropriate to deal with the dispute that arises. The JCT provides principally for the appointment of an adjudicator once notice of intention to refer a dispute has been given, but it also proposes those amendments which are necessary where the parties decide to name an adjudicator in the contract. The appointment of an adjudicator only after the dispute is known has merit but may mean that a number of different adjudicators are appointed during the progress of the works.

Where an adjudicator is not named in the contract the process for securing an appointment is either by agreement of the parties or by application of either party to the nominator specified in the appendix to the contract. Whichever route is used, the appointment of and referral to the adjudicator should occur within seven days of the notice of intention to refer. This raises a question. If an adjudicator is not appointed within seven days of the dispute notice, would this be invalid for the Act? Clause 41A.2.2 is ambiguous. The agreement between the parties with the 'object of securing' the appointment and referral may be distinct from the appointment and referral itself. The words used in the contract could be interpreted differently from those in the Act because one clearly refers to a 'timetable' and the other to an 'agreement' for securing appointment and referral. Such a nuance is best ignored but whether it will be remains to be seen (this is discussed further in respect of clause 41A.4.1). Furthermore, the proposed adjudicator must be willing to enter the JCT standard agreement for adjudicators and this must be executed by the parties and adjudicator (41A.2.3).

The appendix to the contract provides two separate lists of nominators, one for adjudication and one for arbitration, and the parties should identify one by deleting all the others. The president or vice-president of the RIBA is the appointor if nobody else is identified. The JCT has assurances from the nominating bodies listed that they will comply with the timetable for appointment and will only appoint adjudicators willing to enter the standard adjudication agreement.

Clause 41A.3 recognises that an adjudicator may become unavailable to act in the adjudication and provides for a replacement to be named in accordance with the procedures. The adjudicator can be appointed either by agreement or by application to the specified nominating body.

If the parties decide to name the adjudicator in the contract they will need to make the necessary amendments to clause 41A as set out in the guidance notes to amendment 18. This will include an additional appendix entry for the insertion of the adjudicator's name, address and fee. The revised clause 41A.2 states that an adjudicator cannot be an 'employee of or otherwise engaged by either party' unless the parties have agreed. Where the appointment is made at the time of the dispute the status of the adjudicator is known and it is assumed that it will remain the same for the duration of the dispute; but, of course, it may not.

The JCT Adjudication Agreement is published in two forms for use with adjudicators, unnamed or named in the contract. The two are broadly similar and straightforward, providing for the setting out of the names of the parties and the adjudicator, together with their addresses. They also provide space to briefly describe the works and the form of contract. The agreement has five clauses covering:

- appointment and acceptance
- adjudication provisions
- adjudicator's fee and reasonable expenses
- unavailability of adjudicator to act on the referral
- termination.

The agreement refers to the adjudication provisions within the contract, and the adjudicator therefore is obliged to carry out the adjudication in accordance with them. The parties are jointly and severally liable for the fees and reasonable expenses incurred. Where the adjudicator is unavailable to act, a notice must be sent to the parties but although this is part of the adjudicator's agreement there could be many instances where this is not done, either because it is impractical or simply because the adjudicator fails to do so.

Parties may, where they agree, terminate the adjudicator's appointment at any time and the adjudicator will be entitled to a fee and expenses reasonably incurred. However, if the termination occurs because of a failure on the part of the adjudicator, no fee and no expenses are payable. The schedule to the agreement will state the fee to be paid and this may be either a lump sum or based on an hourly rate.

The adjudication agreement for a named adjudicator has some small differences: it makes it an obligation for the adjudicator to 'use ... best endeavours to be available...' and to agree, if asked, to execute a similar agreement in respect of nominated or named subcontract disputes. Another noticeable point concerns the ability of the parties to terminate not only the adjudication agreement but also the adjudication (5.1.2).

The adjudication

Clause 41A.4.1 provides for two stages in bringing about an adjudication, although there seems no reason why these cannot be simultaneous where the adjudicator is named in the contract. It would be problematic where the adjudicator is not named. The first stage is the notice of intention to refer a dispute and the second stage is the need, if one is going to commence adjudication, to refer the dispute within seven days of this notice or the execution of the adjudication agreement if this is later. Clause 41A.2.2 mentions the referral of the dispute to the adjudicator within seven days of the notice of intention to refer, but this is in the context of the agreement (discussed above). However, clause 41A.4.1 states that the dispute can be referred within seven days of this notice or the execution of the adjudication agreement. This latter date could have the effect of delaying the reference substantially and raises the question as to whether it accords with the HGRCA. The contract seems to suggest that a timetable that complies with the Act is provided but where an adjudication agreement is not executed the whole process can be extended. It would seem that the way the provisions have been drafted leaves an element of doubt as to their validity in compliance with the HGRCA.

The referral must set out the particulars of the dispute and include any material that the adjudicator is to consider, together with the basis of the party's arguments and the remedy sought. A copy of all this documentation should be sent to the other party but there is no requirement to send it to the contract administrator. The adjudicator may wish to send this information to the contract administrator and interview him. Whether to copy-in the contract administrator was considered by the JCT because it was felt that it might be worthwhile to provide the contract administrator with the chance to change a matter related to a certificate or the like. In the published amendment, there is no mention of the contract administrator having this facility because the JCT was advised that it would run counter to the HGRCA. A novel attempt to get over this problem was devised by the ICE, with the concept of 'dissatisfaction not constituting a dispute', but whether this will hold up if challenged is another matter.

Transmission of documents is covered in clause 41A.4.2 and includes most recognised means of delivery. It also covers the need to confirm fax transmissions and the deemed date for the receipt by registered post or recorded delivery, which is 48 hours after the date of posting, excluding Sundays and public holidays. This deemed date applies unless there is proof that the actual delivery date is different. The latter provision is not entirely clear in its purpose because the adjudicator is, in any event, required to confirm the date of receipt. Presumably it acts as a default position but if this is the case, the clause could be drafted differently.

The answering party has seven days from the referral to make a response to the adjudicator. This response can include arguments relied on to defeat the claims made by the referring party and any other documents to be considered by the adjudicator.

Under this contract the adjudicator is acting for the purposes of section 108 of the HGRCA and is neither an expert nor arbitrator. This removes any doubt (even if excessively cautious) that the adjudication may subsequently be construed as an arbitrator.

During the adjudication the adjudicator must act impartially but has total discretion as to the procedure to be followed. As the adjudicator has this discretion the inclusion of specified things that can be done is not essential. Nevertheless, the contract includes the ability to require parties to carry out tests, open up the works, visit the site or any workshops and to ask for other information. Clause 41A.5.5.6 allows the adjudicator to secure information from any employee or representative of the parties, but only with prior notice to the relevant party. The purpose of the notice could be questioned because there is no right of objection.

Not all disputes will be resolvable by the adjudicator on his own and consequently he may get technical or legal advice as appropriate, so long as the parties are given notice of this and an estimate of cost. The need to give notice together with an estimate of cost is sensible. It would be imprudent for the adjudicator to proceed with obtaining such advice if the parties did not agree. What happens if the adjudicator believes such help is necessary but the parties say they are unhappy or unwilling to incur the costs? It would seem that as the adjudicator is not bound to obtain agreement from the parties, he could proceed without such agreement if such help was considered necessary, and still recover the costs. Parties to the dispute pay their own costs but the adjudicator may direct who pays the cost of testing or opening up if required by the adjudicator under clause 41A.5.5.4.

Adjudicator's decision

The periods for reaching a decision are as set out in sections 108(c) and (d) and are discussed in Chapter 9. A decision of the adjudicator can take many forms and may be the result of the specific power given to open up, review and revise any certificate, opinion, decision, requirement or notice given under the contract. The decision may include any applicable interest payment relevant to the dispute but only simple interest is payable, which differs from some other adjudication rules. The decision must be given in writing to the parties and no reasons need be provided, nor can they be demanded. Here again the decision need not be communicated to the contract administrator but there is surely no reason why it should not be. Although the adjudicator has very limited powers to direct the parties' costs, the adjudicator must set out in the decision the apportionment of the adjudication fee and expenses. If the adjudicator fails to do so, the fee and expenses are borne equally by the parties. The guidance note states that the parties may submit sealed letters to the adjudicator to be opened after the decision is reached, if there are matters that they wish to be considered before the fees are apportioned. Where one party feels that the other is being unreasonable in considering any offer that may have been made, this is a useful tactic to try to reduce the fees paid.

Decisions given by the adjudicator are binding until the dispute is finally determined either in arbitration or legal proceedings as appropriate. Either party may wish to pursue the matter but the JCT emphasises that this is not an appeal against the adjudicator's decision but rather a new hearing of the dispute as if no decision had been made. In the meantime, the parties must comply with an adjudicator's decision and failure to do so means the party wishing to enforce the decision may move directly to legal proceedings to secure enforcement. Enforcement of a decision is not a matter for arbitration even where an arbitration agreement is operative, as it is specifically excluded in order to prevent a stay of proceedings pending arbitration and to prevent the removal of any right to summary judgment, interim payment or an injunction.

The adjudicator must give a decision and to say that he is unable to come to a decision would not constitute a decision. Furthermore, unless the contract specifically provides it, the adjudicator has no power to direct as part of the decision that money should be held by a stakeholder – a common resort under the old subcontracts.

The adjudicator is given immunity under clause 41A.8 to reflect section 108(4) of the HGRCA but this immunity can only extend as far as actions by the parties themselves.

Amendment 12 JCT Standard Form of Building Contract With Contractor's Design 1981

Amendment 12 to this form includes amendments arising out of the proposals contained within *Constructing the Team* and to comply with the

HGRCA. It was published in April 1998. This amendment is similar to that for JCT 80 as the tribunal has attempted to bring these forms into line regarding adjudication. However, there are some important differences related to the supplementary provisions that were issued in February 1988 as amendment 3. The HGCRA caused the JCT not only to reconsider the main provisions of JCT 81 but also to delete the supplementary provisions relating to adjudication. The supplementary conditions already provided for adjudication in advance of the Act but the statutory requirement for adjudication meant that the main form was revised and that the optional provisions were no longer apposite.

In other respects, amendment 12 with regard to adjudication is identical to the adjudication provisions discussed above in amendment 18 of JCT 80, except for the clause numbering and the new articles 6A and 6B. Clause 39 deals with the settlement of disputes. Additionally a number of different consequential amendments are necessary to the appendix to the conditions of contract.

Amendment 12 JCT Intermediate Form of Building Contract 1984 Edition

The changes made to the Intermediate Form of Contract (IFC 84) in respect of dispute resolution are broadly in line with the JCT 80 changes discussed above, and the adjudication provisions themselves are the same. The difference lies in the fact that the IFC 84 makes no reference to a choice between arbitration and litigation following adjudication. IFC 84 provides, strangely, only for arbitration to follow adjudication and consequently the clause provisions and the articles vary from those in JCT 80 and JCT 81.

The existing clause 9 under the heading of 'settlement of disputes – arbitration' has been deleted. A new section entitled 'settlement of disputes – adjudication – arbitration' has been provided and a new clause 9A dealing with adjudication and a new clause 9B dealing with arbitration have been drafted. These provisions, although making reference to other clause numbers, otherwise mirror those in JCT 80.

Just as the provisions have been amended, the articles have also been changed. A new article 5 has been inserted and this provides that any dispute or difference arising under the contract may be referred to adjudication at any time in accordance with clause 9A. The existing article 5 has been deleted and a new article (article 7) provides that disputes are referred to arbitration in accordance with clause 9B. However, arbitration is subject to the right under article 5 to go first to adjudication if so desired. There are exceptions to the right to arbitrate and disputes in respect of value added tax and the statutory tax deduction scheme are specifically excluded, as is any dispute in connection with the enforcement of any decision of an adjudicator. This last exclusion ensures that the right to seek summary enforcement of the decision is preserved and it avoids proceedings being stayed on account of there being an arbitration agreement. This article contains ten lines without punctuation and gives rise to serious concerns about how well such drafting communicates the intention. It is not difficult to put a number of constructions on the words contained in this article.

Amendment 11 Agreement for Minor Building Works 1980 Edition

The changes made to the minor works form of contract in respect of dispute resolution mirror those contained in IFC 84 as discussed above, but with a number of differences. The minor works form, like IFC 84, provides no choice between arbitration and litigation following adjudication. However, it differs in that neither the amendment nor the guidance notes refer to naming an adjudicator in the contract. This is in contrast to all the other JCT forms previously discussed. Presumably for minor building works the JCT consider it unwarranted to make such a provision; the time and cost are not justified. However, there is nothing to prevent parties from naming an adjudicator if desired and the guidance notes for JCT 80, JCT 81 or IFC 84 would assist. The article numbering varies as does their format. Furthermore, clause 9.0 simply makes reference to the adjudication and arbitration provisions that are included in part D and part E of the supplementary memorandum to the agreement, rather than including all the detail within the clause.

Subcontract forms of contract

The development of dispute resolution in subcontracts is somewhat different, no doubt owing to the greater frequency of disputes in these contracts, especially over payment. The story usefully starts with the subcontract forms used in connection with JCT 63.

The subcontract form, known as the 'Green Form', for use where the subcontractor was nominated under the Standard Form of Building Contract (1963), was also published in 1963 but by the National Federation of Building Trades Employers (NFBTE).

This subcontract contained an arbitration clause but also included a brief set-off clause (clause 13), which enabled the contractor to set off against money due any sums that the subcontractor was liable to pay the contractor under the subcontract. This clause and the resulting litigation led in February 1976 to the introduction of an adjudicator in clause 13B, by way of an amendment to the then current 1975 edition. Similar but not identical provisions (clauses 15 and 16) existed in the subcontract form, known as the 'Blue Form', first published in 1971 also by the NFBTE. The 'Blue Form' was for use where the subcontractor was *not* nominated. A similar amendment was made to the 'Blue Form' at the same time.

The introduction of the revised clause 13 into the 'Green Form' followed a

series of legal cases (*Dawnays* v. *Minter*, 1971; *Gilbert-Ash (Northern) Ltd* v.
Modern Engineering (Bristol) Ltd, 1973; *Mottram Consultants* v. *Bernard Sunley
& Sons Ltd*, 1974; *Kilby and Gayford Ltd* v. *Selincourt Ltd*, 1973) concerning the
sanctity from set-off of certified payments and an employer to set off sums
against a contractor. The recognition that certified sums were not special
encouraged the development of adjudication. An arbitrator could have dealt
with such matters, but it was perceived that arbitration was generally a
lengthy process. With cash flow important, time was of the essence in
arriving at a solution and adjudication was seen as the best means of fulfilling this requirement. In principle there is no reason why arbitration could
not be equally speedy but current attitudes towards arbitration procedures
somewhat inhibit speed.

The new clause 13 in the 'Green Form' exceeded two pages and was the
first attempt under a contract associated with a JCT Form to provide an
interim means of settling amounts of set-off which were not agreed. The
February 1976 amendment substantially replaced the old clause 13 and also
introduced clause 13A to cover two distinct situations regarding set-off.
Clause 13A(1) dealt simply with set-off of agreed amounts, with clause
13A(2) dealing with the more complex situation of set-off where amounts
were not agreed. The latter clause enabled the contractor to set off amounts
as long as the three expressly stated conditions had been fulfilled. Clause
13A(3) provided that any set-off under this clause was without prejudice to
what happened in any subsequent negotiations, arbitration proceedings or
litigation. Clause 13B provided for the appointment of an adjudicator where
the subcontractor wished to challenge the amount of set-off. Any action
under clause 13(A)(2) did not prejudice a further claim under the provisions,
if and when a further sum became due (clause 13(B)(7)).

Once the contractor had notified the subcontractor of the intended set-off,
the latter had 14 days from receipt of the notification to send a written
statement setting out reasons for not agreeing the amount. This written
statement could also include a counterclaim, which was a detailed quanti-
fication. At the same time, the subcontractor was to give notice of arbitration
to the contractor and request action by the Adjudicator named in part xiii of
the appendix to the 1976 edition of the subcontract.

The adjudication clause was to ensure that disputed claims over set-off
could be quickly and independently resolved, without waiting until completion of the project. However, enforceability of the adjudicator's decision
soon became a problem and furthermore, the contractor's notice specifying
the amount and any subcontractor's counterclaim was not binding in any
subsequent pleadings in arbitration. In other words, the use of an adjudicator was an attempt to ensure that appropriate cash flow was achieved in
the context of the various contractual obligations.

It was essential that the adjudicator was named prior to entering into the
contract, otherwise any benefits that might flow would likely disappear. If
not named, problems of agreeing an adjudicator could slow down resolution
of the actual dispute.

Soon after the introduction of adjudication, the NFBTE compiled a list of those willing to be approached to act as adjudicators under clause 13B. In the event that the named adjudicator could not act, the named person could appoint another person so long as that person did not have an interest in the subcontract or main contract or any other contracts with the organisations concerned.

Once the notice for triggering the appointment procedure was given, the contractor could, within 14 days of the receipt of the subcontractor's statement, send to the adjudicator a written statement setting out brief particulars of his defence to any counterclaim made by the subcontractor. Where the contractor provided a written statement, the adjudicator was required to provide a decision within seven days of receiving the statement. If the contractor did not provide a written statement, it would appear on strict interpretation that a decision had to be given on expiry of 14 days from the subcontractor's notice but with no time limit to provide the decision. It seems unlikely that this was intended and seems that it was the intention for the seven days to run consecutively with the 14 day period.

The decision of the adjudicator was to be made without any further statements, unless he was required to explain any ambiguity, and without any personal hearing. The process was to provide a swift dispensation of a fair and reasonable decision. The adjudicator's discretion was absolute and he had no obligation to give reasons. The decision was binding on both parties until the matter was either settled by agreement or decided by an arbitrator or the court. Therefore, the adjudicator's decision could be reconsidered and the arbitrator could, with absolute discretion, at any time before the final award and on application of either party, vary or cancel the decision if it was just and reasonable to do so. It can be deduced from this that the adjudicator's decision was in all likelihood to be only temporary in nature where significant sums of money were at stake.

Although the adjudicator had absolute discretion, he had to act in a manner 'in all the circumstances of the dispute to be fair, reasonable and necessary...'. It was essential for the adjudicator to follow strictly the procedural requirements of clause 13. Failure to do so could result in the adjudicator being sued for damages for negligence by either party[2].

The decisions an adjudicator could make are set out in clause 13B(3)(a) with the disputed amount dealt with as follows:

- retained by the contractor
- pending arbitration, held by a trustee stakeholder
- paid in whole or part to the subcontractor.

The range of decisions was later extended in other subcontract forms (e.g. NSC/4 clause 24.3.1.4) to include a combination of the above options, thus enabling the adjudicator to have available a full range of possibilities. Before this extension, only one of the options could be used in respect of any one dispute.

A trustee-stakeholder was introduced at the same time. Like the adjudi-cator, the trustee-stakeholder was to be named in the appendix. Clause 13B(5) sets out the status and obligations of the trustee-stakeholder. The trustee-stakeholder position was introduced to ensure security of the monies pending later settlement by arbitration. Although monies held attracted interest, one or other or possibly both parties could be denied cash flow as a consequence, with a main contractor perhaps prejudiced by a sub-contractor's later insolvency.

The relationship of the trustee-stakeholder and indeed, the adjudicator to the parties to the subcontract is not governed by the subcontract. Therefore, any remedy for default on their part must be sought under the contract between the contractor and/or subcontractor and the adjudicator or trustee-stakeholder, or in tort or in both contract and tort.

As with any process for resolving differences, there is a cost to be met. The trustee-stakeholder can deduct reasonable and proper charges from amounts held, whereas the adjudicator's fee is paid for by the subcontractor. However, liability for both amounts can be subsequently determined by the arbitrator if the matter ultimately proceeds to arbitration.

The first edition of the 'Blue Form' domestic subcontract (1971) contained a clause (clause 16) allowing for a contractor's right to deduction or set-off in similar but not identical terms to those of clause 13 of the 'Green Form'. The 'Blue Form' introduced new clauses (clauses 15 and 16) which paralleled the new clause 13 of the 'Green Form', except that under the 'Green Form' and where the subcontractor was nominated, any set-off relating to delay could not be made unless the architect had issued the appropriate certificate with a copy to the subcontractor. This meant that the adjudicator under the 'Green Form' but not the 'Blue Form' had to establish that there has been com-pliance with a further procedural requirement. The adjudicator under the 'Green Form' was only required to consider delay once the architect certified that the subcontract works should have been completed by the specified date. There was no such obligation under the 'Blue Form'.

The two subcontract forms adopted adjudication but only for disputes related to set-off. Strangely, adjudication was not extended to other areas of dispute.

Subcontract forms of contract for nominated subcontractors

Amendment 10 of the Standard Form of Contract 1980 Edition, issued March 1991, introduced a revised procedure for the nomination of subcontractors, and the earlier documents and subcontracts were superseded. Under the 1991 procedure, a single method replaced the 'basic' and 'alternative' methods and a new tender NSC/1 form of agreement and subcontract (NSC/A, NSC/C) were introduced.

The agreement NSC/A is separate from the subcontract form and article 4 deals with the settlement of disputes. It broadly replicates NSC/4 (its predecessor) in that all disputes or differences are to be referred to arbitration. Similarly, NSC/C provisions for set-off and the appointment of an adjudicator are the same as those in NSC/4 and bear a resemblance to the Green Form provisions. The relevant provisions are found in clauses 4.26–4.37, now deleted under amendment 7 (1998).

The names of the adjudicator and the trustee-stakeholder are to be inserted in item 3 of part 3 of the tender NSC/T form. Item 7 of part 3 allows the parties to identify the body which will appoint the arbitrator and where no appointor is selected, the president or a vice-president of the Royal Institution of Chartered Surveyors is adopted.

JCT issued amendments to the nominated subcontract conditions in April 1998 to take account of the HGCRA and to reflect the changes to the main form. The existing article 4 has been replaced with the following words: 'If any dispute or difference arises under this Sub-Contract either party may refer it to adjudication in accordance with clause 9A.'

Except for the reference to subcontract instead of contract, this article follows the wording of JCT 80 amendment 18. Similarly, additional articles, numbered 5A and 5B, have been included to cover the reference of disputes to either arbitration or litigation.

These articles have the same effect as articles 7A and 7B of amendment 18. Amendments have been made to the NSC/T; item 12 states that the option selected in the main contract will also apply to the subcontract unless otherwise agreed. This seems sensible, but more significant is part 3 of the NSC/3 which identifies which of these is to apply. If the parties desire arbitration they add nothing to the entry in part 3 as arbitration is the default forum unless litigation is specifically chosen. Litigation is chosen by the deletion of the words 'clause 9B (arbitration) applies'.

Similarly, the body for nominating the adjudicator will be the same as that for the main contract unless otherwise agreed. The appropriate selection is made in part 3 and this should preferably correspond to the main contract choice.

Section 9 of the NSC/C subcontract conditions, as amended by amendment 7, provides for the settlement of disputes and except for references to subcontract and its clauses, it follows the same wording as the amended clause 41 of JCT 80. It also provides for the adjudicator to be appointed once a dispute arises but provides for the alternative of naming an adjudicator within the subcontract. The guidance notes make no comment on whether it is appropriate to appoint the same adjudicator on both a main contract and a subcontract, but this is clearly a matter for the parties to decide. However, if the adjudicator is named within the contract, the adjudication agreement places an obligation on him to act in relation to a dispute arising under a nominated or named subcontract if required to do so. The words used in the JCT adjudication agreement (named) are the same for both the main and subcontracts and therefore the latter could also apply to disputes with nominated and named subcontracts (if such existed) within the subcontract.

Standard Form of Employer/Nominated Subcontractor Agreement NSC/W 1991 Edition

Amendment 2 was issued in April 1998 and, among other things, sets out the changes in respect of dispute resolution. These also follow the general pattern adopted by JCT 80. Clause 11.1 is new and provides for either party to refer a dispute to adjudication. This is in addition to the right to pursue the dispute through arbitration or in legal proceedings, whichever applies. There is no choice of arbitration or legal proceedings under the NSC/W because the warranty simply follows whatever applies under the main contract. Even where arbitration is applicable, enforcement of an adjudicator's decision is for litigation and does not fall within the scope of the arbitration clause.

Clause 11.2 states that the law of England will apply, whatever method is used for settling disputes. Clauses 11A on adjudication, 11B on arbitration and 11C on legal proceedings all mirror the main contract.

NAM/SC subcontract conditions

The NAM/SC subcontract conditions for use with IFC 84 were amended to comply with the HGCRA by amendment 11. The amendments follow the same lines as those for IFC 84. The option of legal proceedings in this form is again precluded. The amendment does not refer to the possible naming of the adjudicator in the subcontract. This is not that surprising in that the scale of works is likely to be small and the JCT have adopted an approach consistent with that in the minor works form.

Standard Form of Subcontract for Domestic Subcontractors (DOM/1–1980 Edition)

This subcontract is published by the Building Employers Confederation for specific use with domestic subcontractors appointed under clauses 19.2 and 19.3 of the JCT Standard Form of Building Contract 1980. Its relationship to the modern main form of contract is therefore similar to that of the 'Blue Form' subcontract to JCT 63.

Although the relationship is similar, the DOM/1 subcontract ties in more closely with the general approach adopted by JCT. DOM/1 therefore deals with the settlement of disputes in an identical way to that found in NSC/4, NSC/4A and NSC/C. Article 3 requires that any dispute or difference, except for specific disputes under the Finance Act 1975 Tax Deduction Scheme, shall be referred to arbitration in accordance with clause 38, regardless of when they occur, subject of course to the Limitation Acts. Part 1 section B of DOM/1 refers to the main contract and the reference here is to the appointment of the arbitrator to settle disputes between the employer and the contractor. The appointment of the arbitrator to settle disputes between the subcontractor and the contractor is dealt with in part 14 and adopts the same default position as the NSC/C where no appointing body is identified.

The contractor's right of set-off and associated provisions are contained in clauses 23 and 24 of DOM/1 and are broadly the same as those in the nominated subcontract form, except for three main procedural differences:

- no reference is made to the fiduciary obligation of the contractor under clause 21.9.1,
- an architect's certificate of delay is not a requirement before a set-off can be made for loss and expense associated with delay in completion,
- there is no requirement of the subcontractor to provide a copy of the architect's certificate certifying delay.

This subcontract provides for the adjudicator to be named in part 8 of section C of the appendix contained within the articles of agreement. The contractor or subcontractor can suggest the name to be inserted, but either way it should only be following agreement with the other party. Part 8 of section C of the appendix also provides for the naming of the trustee-stakeholder.

DOM/1 1998 revision

When DOM/1 was revised to take account of the HGCRA 1996 it was far from clear what had changed. DOM/1 comes in two parts, the articles of agreement and the subcontract conditions, and reprinted versions of these were issued in May 1998. The articles of agreement incorporated a number of changes referred to later, but the front of the contract stated that this was the 1980 Edition incorporating (among others) amendment 10, which was to cover the changes necessary to comply with the HGCRA. The subcontract conditions, however, had no such reference on the front cover although they contained a number of changes, including dispute resolution. In July 1998, a correction sheet was issued which indicated a large number of alterations, but the most pertinent for present purposes was that the reference to amendment 10 should be deleted from the front cover of the articles. This at least made the articles consistent with the conditions but left a subcontract which simply appears to be a 1998 reprint. This is not the case as it is far more than just a reprint. Perhaps it is not surprising that both bodies representing subcontractors, who approved the earlier contract, no longer approve the 1998 form. This is apparent from a correction sheet issued to article 1.3 omitting the names of the National Specialist Contractors Council and the Specialist Engineering Contractors Group.

Articles of agreement 1998

The articles have been amended appreciably with the existing article 3 replaced with articles 3 and 4 which deal respectively with adjudication and the settlement of disputes by arbitration or litigation. The two new articles follow the broad principles of JCT 80 but in the subcontract form they are combined and the words are not identical even after making the necessary adjustments for the fact that it relates to the subcontract. Both parties have the right to refer a dispute or difference to adjudication and, subject to this right, they may also refer the matter to arbitration or legal proceedings whichever is applicable. Which is applicable depends on the entry in the appendix to the subcontract. The important correction issued in July 1998 now refers correctly to the subcontract clause reference instead of the main contract clause. There are exceptions to the matters that may be referred to arbitration, the most important being the enforcement of an adjudicator's decision. Although the intention of the contract, when read together with the JCT form, is clear, there could be some confusion on this point because the articles are badly drafted.

The appendix to the articles incorporates the JCT 80 appendix which itself incorporates changes to clause 41 dealing with main contract disputes. Section C part 8 provides for the parties to select the body to nominate the adjudicator and where no body is identified the president or vice-president of the Royal Institution of Chartered Surveyors will make the appointment. This part also refers to 'clause 38 applies' but this should state 'clause 38B applies' (as indicated in the corrections), with arbitration the chosen forum unless this is deleted. It needs to be deleted if legal proceedings are to apply instead of arbitration. The intention is the same as that for JCT 80 but was poorly accomplished. Indeed, it probably was not achieved until the correction was issued. Part 14 makes provision for the selection of an arbitrator by an appointing body in a similar way to that of the adjudicator.

Subcontract conditions

It is most surprising that the 1998 reprint does not refer to the amendments incorporated because clause 38 has been significantly changed along the same lines as JCT 80 clause 41. It provides for the settlement of disputes by adjudication, arbitration and legal proceedings. Clause 38A sets out the adjudication provisions and clause 38B covers arbitration; apart from the different clause references they are as in JCT 80. The adjudication provisions are self-sufficient and no further adjudication rules are required. The arbitration provisions contain a number of other differences such as:

- Clause 38B.2 which states that 'In any such arbitration as is provided for in clause 38B any decision of the Architect which is final and binding on the Contractor under the Main contract shall also be and be deemed to be final and binding between and upon the Contractor and the Sub-Contractor'.
- Clause 38B.3 where the arbitrator is given powers to open up and review a range of matters but with the notable exception of '...a decision of the architect to issue instructions pursuant to clause 8.4.1 of the main contract

conditions and which instructions were issued to the sub-contractor pursuant to clause 4.2.2...'
- Clause 38C makes provision for a dispute or difference to be determined by legal proceedings but this will only apply if the appropriate deletion of the reference to the arbitration clause in part 8 of the appendix has been made.

Furthermore, the contractor's right to set-off and associated matters contained in clauses 23 and 24 of the earlier version of DOM/1 are omitted. Thus one of the first uses of adjudication in construction, set-off, is superseded by the general right to adjudication for all disputes and differences.

Chapter 11

General Conditions of Government Contracts for Building and Civil Engineering Works

Introduction

This chapter considers the general conditions of contract devised for use by government departments when carrying out building and civil engineering works. A brief introduction to the first and second editions is followed by a more detailed commentary on the third edition and GC/Works/1 (1998).

Form GC/Works/1 Editions 1 and 2

The first edition of this contract, published in November 1973, contained an arbitration clause but no clauses related to alternative dispute resolution or an adjudicator. Clause 61 provided for 'All disputes, differences or questions between the parties to the Contract with respect to any matter or thing arising out of or relating to the Contract', but with some notable and important exceptions. The exceptions are referred to as '... other than a matter or thing arising out of or relating to Condition 51 or as to which the decision or report of the Authority or of any other person is by the Contract expressed to be final and conclusive ...' Many issues fell to be determined by the supervising officer and/or authority, with the decision to be final and conclusive. Therefore, the impact of the arbitration provisions was restricted in their application.

In September 1977, the second edition of the GC/Works/1 was published. The arbitration provisions were identical. This edition was superseded by GC/Works/1 Edition 3 in 1989 with a revised version, published in 1990.

Form GC/Works/1 Edition 3 (Revised 1990)

Generally

This contract was derived from the earlier editions but incorporated many changes. In addition, GC/Works/1 Edition 3 adopted a different format, and reference to the section entitled 'Particular Powers and Remedies' quickly revealed an important new provision – adjudication. The inclusion of this provision in clause 59 was a significant development and this together with a new arbitration clause (60) provided a very different framework for dispute resolution.

Arbitration

Clause 60 provides that in addition to adjudication, arbitration will be available. The provisions cover all disputes relating to the contract but, as in earlier editions, with the exception of a decision expressed to be final and conclusive, which could not be opened up. Unless otherwise agreed, arbitration cannot commence until after completion, alleged completion, abandonment of the works or determination of the contract.

Adjudication

Adjudication is contained in clause 59 which incorporates a number of revisions from the 1989 version of GC/Works1 Edition 3, including changes to three subclauses. Adjudication was provided as an alternative way of resolving disputes and not as a condition precedent to arbitration. The decision of the adjudicator is final and binding on any matter referred, until completion, alleged completion, abandonment of the works or determination of the contract. At any one of these points an arbitration can be commenced and the adjudicator's decision challenged because arbitration is 'In addition to adjudication . . .' Nevertheless, a formal mechanism has been established that will enable disputes to be addressed earlier than might otherwise be the case and possibly resolved to the satisfaction of the parties, thus obviating the subsequent need for arbitration. In any event it provides a temporary decision that is binding while work is continuing. Technically, because of the words used in subclause 7, once completion is reached the adjudicator's decision ceases to be binding and can be reviewed in any way. But, in practice, a decision will only generally be challenged by reference to arbitration. The 1989 edition used different words in subclause 7 and did not refer to the binding nature of the adjudicator's decision, only that such a decision could not be questioned until after completion.

The operation of the adjudication provisions cannot be instigated until a dispute has been outstanding for at least three months (clause 59(1)). This is a rather strange restriction in that it prevents the use of an adjudicator at the stage when it is likely to be most beneficial. Furthermore, the date when the dispute first appeared is also problematic in that no formal notice is required for the purposes of this clause. The first formal notice is given by the contractor when adjudication is sought and, strictly, this is only valid provided the dispute has been outstanding for three months. There are instances under the contract where the contractor is required to act when dissatisfied with a decision. These could identify the date when a dispute commenced but even this is far from clear. For example, under clause 36(5) if the contractor is dissatisfied with any decision in respect of an extension of time, a claim must be submitted within 14 days of the decision.

Adjudication is available for 'any dispute, difference or question arising out of, or relating to, the Contract during the course of the Works' except for a decision expressed to be final and conclusive. This provision has the same restriction as the arbitration clause in that it cannot operate when the decision is said to be final and conclusive. This is a major change from the 1989 edition which did not make it explicit that decisions said to be final and conclusive were outside the adjudication procedures. Consequently, the new wording makes it explicit that disputes concerning the following matters cannot be referred, because the Authority's decision on these matters is final and conclusive:

- clause 6(1) – cessation of the employment of any person
- clause 18(3) – QS measurements taken in absence of contractor
- clause 24(3) – matters of corruption
- clause 26(3) – admission of persons to site
- clause 31(6) – fitness and suitability of any thing to be incorporated
- clause 37(7) – taking of early possession
- clause 39(2) – issuing of completion certificates
- clause 40(3) – issuing of instructions considered necessary or expedient
- clause 44(5) – adjustment for labour tax matters
- clause 50(3) – certification of payment except for the final sum due to the contractor
- clause 56 – matters related to determination
- clause 63 – concerning objections to subcontractors or suppliers not provided for in the documents.

Whether the exclusion of all these matters leads to a balanced contract is questionable in that it denies an opportunity to resolve a dispute rather than simply have a decision imposed. The apparent justification on matters such as those relating to the employment of persons on site is understandable, but it is far more difficult to see the logic in excluding matters related to the certification of payment.

Furthermore, the adjudicator can only be used for disputes 'arising out of, or relating to, the Contract during the course of the Works'. Once the works have been completed, the only dispute forum is arbitration. But what if disputes arise during the course of the works where the adjudication provisions are not operated prior to completion? It would seem that the intention of the contract was probably to submit the dispute to arbitration but this again is uncertain because the words used are capable of different interpretations.

It is clear that although the provision of adjudication is a useful addition to the contract, the severe curtailment of its operation and the appointment of the adjudicator by the client could substantially defeat the objective of an 'alternative' approach to dispute resolution.

In any adjudication provisions, it is important to specify the procedural arrangements, either in detail or by reference to a particular scheme. This contract does not identify a scheme but spells out in clause 59 the operation of the adjudication provisions. An adjudication can be commenced when the contractor gives notice of a dispute to the person named in the abstract of particulars. For the sake of clarity a provision is needed requiring notice that a dispute has arisen, regardless of whether there is any intention to proceed to adjudication or arbitration. As the notice under clause 59(1) is the first notice referred to in the contract, a party must be certain that a dispute has been outstanding for three months otherwise the notice is invalid and adjudication cannot commence. The practical significance of such a notice may only be that it formally records the existence of a dispute. Strictly, a further notice would be required once the three-month period was fulfilled. In this notice, the contractor must specify the matter in dispute, set out the relevant facts and the main points of argument and provide all relevant documents. Copies must be sent to the project manager and quantity surveyor in addition to the adjudicator.

Once a valid notice has been given, the person named in the abstract of particulars must nominate as adjudicator an officer of the authority or someone acting for the authority who has not been associated with the contract, and must inform the project manager, quantity surveyor and contractor. This may seem unfair to the contractor. Firstly, the person named in the abstract is not the adjudicator but the person who nominates the adjudicator. Secondly, the person nominated can be an officer of the authority and although this person may well act with integrity, the perception held by the contractor is likely to be that this person cannot be truly independent. The nominated person can be, alternatively, someone acting for the authority and although this may be viewed as preferable, true independence may not be achieved. In fact, it could in reality be a worse solution from the contractor's point of view.

Under 59(4) the project manager and quantity surveyor have 14 days from the receipt of the contractor's notice in which to make representations to the adjudicator. What appears as a straightforward provision poses a number of practical difficulties. First, the contractor's notice must be valid, otherwise time does not begin to run. Secondly, the representations have to be made to the adjudicator and one may not have been appointed within this period. There is no time stipulated during which the person named in the abstract must appoint an adjudicator, yet the adjudicator is required to give a decision within 28 days of the contractor's notice. This apparently holds notwithstanding a delay in the appointment of the adjudicator in accordance with clause 59(3). However, it cannot apply if the adjudicator has not been appointed within that period and, as previously mentioned, there is nothing that specifically requires this to happen. This would appear to be an oversight and one which could further slow the process, or worse still fail to bring the adjudication provisions into operation. Without a specified time limit for nominating the adjudicator it is likely that a reasonable time would be implied and presumably this should take into account the need for a decision to be given within the 28 days.

The adjudicator's decision need not provide reasons but it must have regard to how far the parties have complied with any procedures in the contract relevant to the matter in dispute and to what extent each of them has 'acted promptly, reasonably and in good faith'. This could also apply to the time it took in nominating the adjudicator. Clearly, it is in the interests of the parties to act properly otherwise the adjudicator is required to take this into account. The decision must also state how the adjudicator's costs are to be apportioned and failure to do so will mean each party bearing their own.

GC/Works/1 (1998)

Introduction

The latest version of this form is the GC/Works/1 (1998), published in the early part of 1998 in response to Latham's report *Constructing the Team* and the enactment of the HGCRA 1996. This contract was produced by the Property Advisers to the Civil Estate (PACE) who took over the management of these contracts from 1 April 1996.

GC/Works/1 (1998) is a further development of Edition 3 and takes account of the PSA/1 form of contract produced by TBV Consult, published in June 1994. Although GC/Works/1 (1998) is based on earlier editions, it is not to be cited as the fourth edition – an obvious attempt to change the image of government contracts. Although based on earlier editions this contract is very different, being far less authoritarian.

GC/Works/1 (1998) is published in four volumes. These are the with and without quantities editions, the design and build conditions and the model forms and commentary. The adjudication provisions are the same in all three sets of conditions. Reference is made to the model forms published for use in conjunction with the contract. The use of the model forms is not mandatory but they have been carefully prepared and there seems to be every good reason to use them, where applicable. Generally the forms have been drawn up for use by the employer, but where applicable and with suitable amendments can be used by contractors.

Tender, contract agreement and abstract of particulars

GC/Works/1 (1998) was drafted to meet the requirements of the HGCRA 1996 and therefore it is designed for use where the Act applies. Although the Act applies to a wide range of construction contracts, it does not apply to all and the commentary to the form makes it clear that this contract can still be used. However, the commentary also says that where it is considered appropriate to delete the specific provisions required by the Act, a supplementary condition could be used. There is a presumption that the form will not be amended, but if it is legal advice should be taken.

The tender and tender price form, which is defined in clause 1 (1) as a part of the contract, contains in item 11 the following:

'We agree that differences or questions arising out of or relating to the Contract shall be resolved in accordance with Conditions 59 (Adjudication) and 60 (Arbitration and choice of law) of the General Conditions.'

The contract agreement also refers to these clauses but is prefaced with 'Any disputes, differences...'. It is clear that the contract provides for two separate means of settling disputes and differences that may occur.

The abstract of particulars and addendum makes provision for the naming of an adjudicator and the person to be appointed should be stated together with the address. There is also provision for the naming of a substitute adjudicator where the first named is deceased, unwilling or unable to act, or is not independent of either party, the project manager or quantity surveyor. If both named adjudicators cannot act for any of these reasons an appointment will be made by mutual agreement of the parties. If the parties fail to agree then either party may request the president or vice-president of the Chartered Institute of Arbitrators, or the equivalent in Scotland if it is a Scottish contract, to make a nomination. This is also provided for in clause 59(3)(a).

It is the intention to name the adjudicator in the contract but if this is not done, for whatever reason, clause 59(3)(c) applies. Because there is a statutory right to an adjudication the parties should still mutually agree one or, alternatively, resort to the appointment by someone on whom they agree. If the parties fail to do this, or worse still if one party deliberately chooses not to, the adjudication provision would fail and it would seem that the Scheme for Construction Contracts would apply. The specific provisions of clause 59(3)(c) are dealt with later.

Care must be taken too when naming the adjudicator as that person will deal with any disputes regardless of their nature. Subsequently, this may prove to either be an advantage or disadvantage to one party. There is however nothing in principle to prevent the abstract of particulars being extended to provide for a number of adjudicators to cover different types of dispute.

The naming of the adjudicator can be done by the employer and, where this is done, the contractor, in tendering, would accept this person unless the tender was conditional. Alternatively, the naming may be done jointly before, or simultaneously with, the formation of the contract. The prescribed form of adjudicator's appointment is referred to and model form 8 is provided in the documents entitled 'Model Forms and Commentary'.

Similarly, provision is also made in the abstract of particulars for the appointment of an arbitrator.

A footnote to the abstract states that the named adjudicator and arbitrator should be used in all the employer's contracts relating to the project, regardless of whether they are with contractors, consultants or others. This does seem to suggest that the authors do not see naming of adjudicators for particular types of dispute as a serious issue, but it does at least provide for consistency of decisions when dealing with similar disputes.

Appointment of adjudicator where named in the abstract

Where the adjudicators are named in the abstract of particulars, it is prudent to appoint them by entering into the prescribed form of agreement as soon as possible after the main contract has been executed. If this is not done, clause 59(3)(c) requires that the adjudicator shall be appointed within seven days of the notice of intention to refer the dispute to adjudication.

This part of clause 59(3)(c) is straightforward but thereafter the wording is tortuous and the intention lacks clarity. It would appear that if the parties have named the adjudicator(s) in the abstract of particulars at the time of entering the contract but have not appointed them, they must during the 28 days from acceptance of the tender, 'use all reasonable endeavours' to complete an appointment. This applies to the first named adjudicator and the substitute adjudicator.

It would be unusual to have both commenced work and for a dispute to have arisen within 28 days of the acceptance of the tender, but were this to happen a problem could occur in complying with the seven-day timescale, as provided under the contract (and the Act). Consequently, it would be wise to appoint the adjudicator, if not already appointed earlier, before commencing work.

Where the parties fail to complete the required appointments within the 28 days of acceptance of the tender, either party on their own may complete the appointment. This provision may encourage the parties to reach agreement within the time period but the failure to appoint may be down to the adjudicator not entering into the prescribed form of agreement and/or agreeing to the time limits set down in the contract. In these circumstances, it does appear rather strange that a party could then act unilaterally, especially as the parties jointly will, in any event, need to enter into a contract with the adjudicator.

Clause 59(3)(c) states:

> 'If it becomes necessary to substitute as adjudicator a person not named as adjudicator or substitute adjudicator in the Abstract of Particulars – the Employer and Contractor shall jointly proceed to use all reasonable endeavours to appoint the substitute adjudicator – or alternatively to appoint the substitute adjudicator.'

This is not easily understood, but as you can only substitute for someone who is already known it must be intended to cover the situation whereby another adjudicator is required and the person required is not one already named in the abstract. This would accord with the provisions in the abstract where the adjudicator and substitute adjudicators cannot act and a substitute is required.

In circumstances where the parties appoint an adjudicator and substitute adjudicator and they later become unavailable to act, the parties have 28 days from selection of the adjudicator in which jointly to appoint a substitute. The parties choose another adjudicator by mutual agreement but, failing this, the appointment is made by the nominated person (referred to above). Either of these could establish the selection of the adjudicator and time runs from this date. Again, failure to do so means either party can then proceed to appoint on their own but once again the parties jointly will, in any event, still need to enter into a contract with the adjudicator.

If a dispute arose within this 28 day period, this period would presumably not apply as an appointment must be made within seven days of the notice of dispute (as referred to above). Or, does it mean the provisions do not accord with the Act and hence the Scheme would apply, not only to this issue but also to adjudication as a whole?

Clause 59(3)(c) states 'If it becomes necessary to substitute as adjudicator...' rather than the words 'unavailable to act'. This raises an issue as to what event might make substitution necessary. 'Necessary' should be interpreted so as to cover those situations referred to in the abstract, that is, deceased, unwilling or unable to act, is not or ceases to be independent. To interpret this more widely could give rise to haggling over whether another appointment can be or should be made.

The appointment of adjudicators under these provisions shall be under the prescribed form of agreement, 'so far as is reasonably practicable'. This means a different form of agreement could be used in order to secure an appointment, but care should be taken to ensure that the adjudicator is still required to act in accordance with the Act. In particular, it should be noted that under clause 59(3)(b) appointment of the adjudicator is subject to the condition precedent of the adjudicator notifying his agreement to comply with the contractual time limits. Such compliance is provided for in model form 8 but does pose a potential problem when not using this form, as the adjudicator may not then be bound to act in accordance with the contract and its time limits. Clause 59(3)(b) would have no binding effect on an adjudicator unless the adjudicator's agreement so provided.

Once an appointment is made, under clause 59(3)(c), it shall not be amended or replaced without consent of both parties.

Appointment of adjudicator other than by naming in the abstract of particulars

This contract is designed for the adjudicator to be named and consequently it is unwise not to name the adjudicator before or simultaneously with entering the main contract.

If the parties choose not to name the adjudicator in the abstract of particulars, or have failed to do so, at the time of entering the contract they might assume that clause 59(3)(c) provides for this eventuality. However, this clause is concerned with the appointment and substitution of an adjudicator rather than his initial selection. Nevertheless, an adjudicator could be selected and appointed within seven days, as provided under the contract (and the Act), but there is no process specified for achieving this end. There is a timetable, as required by the Act, but whether it is with the 'object of securing appointment' is another matter. It seems unlikely that parties would decide not to name an adjudicator but they may forget. If they do forget the Scheme would probably then apply if the parties could not otherwise agree.

Model form 8

This sets out the agreement to be entered into between the parties to the main contract jointly and the named adjudicator. It provides for the adjudicator to have a copy of the main contract prior to entering the agreement. A copy of this agreement should also be provided for each party.

In England, Wales and Northern Ireland the agreement is to be executed as a deed and requires the adjudicator to act in accordance with the main contract, except where facts or circumstances beyond the adjudicator's reasonable control make this impossible. The adjudicator is required to act in accordance with the adjudication provisions and time limits, set out in condition 59 of the main contract, and must be and remain independent of either party, the project manager and the quantity surveyor. The appointed adjudicator must act impartially and if the adjudicator ceases to be independent, there is a duty to inform both the employer and the contractor. It would be possible to act impartially, even if not be independent, but this is not permitted under clause 59(3)(a). Similarly, an independent adjudicator does not necessarily mean the decision will be impartial and consequently the adjudicator's agreement makes impartiality a separate requirement.

Adjudicators do not have to be the font of all knowledge and they may take other independent legal or professional advice to assist them in arriving at a decision, if reasonably necessary. Any reasonable costs incurred are recoverable from the employer and contractor, together with the adjudicator's fees, expenses and any other costs identified in the schedule. The schedule is referred to in the model form and is for setting out the adjudicator's fees, expenses and any other relevant costs that may be recoverable. Condition 6 also provides for value added tax to be added.

The adjudicator can, where it is reasonable to do so, employ others and charge the parties for the advice without any further reference to them. Where an adjudicator feels the necessity for such help, it would be prudent to inform the parties accordingly before seeking such expertise.

The adjudicator is bound under the agreement, as are any persons who assist, to confidentiality except for the purposes of the agreement itself. All such persons must also comply with the Official Secrets Act 1989 and the provisions of section 11 of the Atomic Energy Act 1946.

Under clause 7 the adjudicator, together with any employee or agent of the adjudicator, is generally free from any liability arising from the performance of their duties, except when acting in bad faith. What constitutes bad faith is open to interpretation but would be based on public law cases and would be a deliberate act to disadvantage.

The proper law of the adjudicator's agreement is the same as that of the main contract. Where the proper law is Scottish law, the model form will require to be amended accordingly.

Adjudication clause

Both GC/Works/1 Edition 3 and the PSA/1 forms of contract provide for adjudication in clause 59. The same clause number is used in GC/Works/1 (1998) but the provision has been substantially expanded.

There is nothing in the contract that specifically states that a reference to adjudication is a condition precedent to arbitration. Because there is a restriction on the ability to refer a dispute to arbitration until after completion, unless otherwise agreed or there is abandonment or termination, adjudication is always available during the progress of the works. Once completion is achieved, it would seem that the contract permits a dispute arising thereafter to be referred directly to arbitration, without agreement, notwithstanding that clause 59(1) enables either party to notify an intention to refer a dispute to adjudication at any time. This means that a reference to arbitration could be made rather than to adjudication, but because both parties have a statutory right to adjudication, if either wishes to exercise that right they could not be prevented from doing so. The question that may emerge is whether the contract's adjudication provisions apply, or whether the Scheme for Construction Contracts should apply. Whether this is an issue depends on tactics in the dispute arising.

Furthermore, once a matter has been referred to an adjudicator and notwithstanding completion of the works, the matter cannot be referred to arbitration until after the expiry of 28 days from receipt of the referral notice or an extension in accordance with clause 59, unless the adjudicator's decision has been given earlier. Once an adjudicator's decision has been given, arbitration can be commenced. Clause 60(2)(b) provides for the three separate events – notification of adjudicator's decision, expiry of 28 days from receipt of referral notice, and expiry of time allowed under clause 59. However, two of these are combined in the drafting and hence correct reference in this clause to 'whichever is earlier' and not 'whichever is earliest'.

Notification

The notification by the employer or contractor of the intention to refer a dispute, difference or question arising under or out of the contract may occur at any time. Notification can be on the 'Employer's Notice of Intention to Refer to Adjudication' model form 20, suitably amended where the contractor is giving notice. A notice can be given in respect of any dispute because the 'final and conclusive' clauses that peppered the earlier editions of the GC/Works 1 have been removed.

Not every notice of intention will lead to the appointment of an adjudicator but, where an adjudicator is required, a further notice needs to be given within seven days of the notice of intention to refer. Such further notice can be by issuing to the adjudicator the 'Employer's Notice of Referral to Adjudication' model form 21, suitably amended where the contractor is making the referral. Clause 59(2) requires this notice to set out the principal facts and arguments relating to the dispute with all relevant documents enclosed. The model form refers to three schedules. The first sets out the dispute, the second the facts and arguments and the third specifies all relevant documents in the possession of the party referring the dispute. Copies of this notice, together with the schedules and copies of relevant documents, are sent to the other party and also to the project manager and quantity surveyor.

Clause 1(3) provides that notices may be given to the contractor by delivery to the contractor's agent, or the employer by delivery to the project manager. Alternatively, the notice may be delivered to the registered office or last known place of business of the contractor or employer, whichever is to receive the notice. Clause 1(4)(b) sets out how timescales are to be calculated and these clauses use the wording from section 116 of the HGCRA. According to clause 1(4)(a), the timescales related to adjudication can be extended by agreement, even though the specified time has elapsed. This raises an interesting point in that such agreement could nullify the periods set out in the Act, but such agreement would not infringe the Act because the contract complied with the Act by the appropriate inclusion of the various requirements. However, such an agreement must be after the dispute has arisen, otherwise it will run counter to section 108(2) of the HGCRA.

The adjudication

The other party, the project manager and the quantity surveyor have seven days from the receipt of the referral notice to submit to the adjudicator anything they feel is relevant to the claim by way of documentation or explanation. Receipt of a notice is taken to be the date that it would arrive in the ordinary course of post. With the vagaries of the postal system, it would be possible for a notice to be deemed to be received much earlier than the actual receipt. This could limit dramatically the already short time for responding to such a notice.

The adjudicator has under clause 59(6) wide discretion as to what procedure is to be adopted and may take the initiative in ascertaining both the facts and the law appertaining to the dispute. The dispute can in effect be decided on a 'documents only' basis, as there is no requirement for the adjudicator to hold a hearing or to visit the site. But if the adjudicator decides that it is desirable for either or both of these to occur, then these can take place. The adjudicator will be influenced by the need to give a proper decision within the time set down.

Clause 59(6) provides that the adjudicator has the powers of an arbitrator acting under clause 60 but, notwithstanding this, acts as an expert adjudicator and not as an arbitrator. What the courts will make of this if it is brought before them is unclear. The adjudicator is said not to be an arbitrator but an expert adjudicator. Will they conclude that this is the same as an expert? Or will they come to the view that this is neither an arbitrator nor an expert. The difference could be important as it determines the framework in which they operate. The powers of an adjudicator under clause 59(8) include the power to vary or overrule any decision made under the contract by the employer, project manger or quantity surveyor, except for the matters listed in this clause. These exceptions exist because by their nature it is generally impractical to go back on such decisions, but the adjudicator still retains the ability to give a decision that provides for financial compensation as the remedy.

Section 108(4) of the HGRCA requires the contract to provide protection to the adjudicator for their actions and omissions and this is stated in clause 59(11) using the wording of the Act.

Adjudicator's decision

The adjudicator is required to notify his decision within 28 days of the receipt of the notice of referral or in such later time as has been agreed by the parties, so long as such agreement is made after the referral of the dispute. The parties themselves may instigate the extended period but more likely it will flow from a request by the adjudicator, and in these circumstances it would be wise for the parties to give it serious consideration. Failure of one party to agree may prove detrimental to their interests.

Clause 59(5) not only provides a maximum time for the decision to be given but also sets out a minimum period of 10 days. This is a procedural matter and ensures that the adjudicator does not make a decision before the other party has had the seven days in which to respond. This clause also provides, as required by the Act, for the adjudicator, with the consent only of the referring party, to extend the 28-day period by up to 14 days. If an adjudicator's decision is given outside the time allowed it will still be valid and this provision overcomes a weakness of the Act.

The format of the decision is for the adjudicator to decide and can be in any form that accords with the adjudicator's powers and embraces the award of damages and legal and other costs and expenses. In addition, the adjudicator can award interest. This may be either simple or compound interest as is appropriate to provide a just decision. Interest may be more generous than the restricted finance charges under clause 47. The interest, as provided for in clause 59(6), is calculated on the whole or part of any amount awarded in respect of any period up to the date of the award and additionally thereafter on outstanding amounts until payment. It also provides that where an amount claimed in the adjudication proceedings is outstanding at the commencement of the adjudication but is subsequently paid before the award, then interest can be applied to these amounts. This ensures that a party cannot withhold monies and later release them without the prospect of having to pay interest on an amount that is wrongly withheld.

The adjudicator's decision must indicate how the adjudication fee will be split and whether one party is to bear any element of the other's legal costs and expenses.

Under clause 59(7) decisions of an adjudicator are binding until a matter is finally determined in legal or arbitration proceedings or in circumstances whereby the parties agree not to accept the decision. Once an adjudication decision has been notified, whether given or not within the time limit, the parties have 56 days from receipt of the decision in which to give notice of arbitration (clause 60(1)). If no such notice is given, the adjudicator's decision becomes unchallengeable and hence final and binding.

The Act does not require the adjudicator to give reasons for his decision but under clause 59(10) either party has 14 days from notification of the decision in which they may request the reasons. Where such a request is made the adjudicator must then state the reasons for his decision.

An adjudicator's decision must be complied with, regardless of any subsequent challenge. The parties are required to comply forthwith, i.e. they must act immediately to give effect to the decision. If a party does not give effect to a decision of an adjudicator the other party can apply for summary judgment and enforcement in the courts. This is provided for in clause 59(9).

PSA/1

This form of contract was referred to above and although not a government form, it is a little used standard form for use in both private and public sectors for UK building and civil engineering works. The PSA/1 form is unlike the GC/Works forms which were intended for government contracts only. The PSA/1 was produced by TBV Consult, in association with a city firm of solicitors, to meet the privatised needs of the Property Services Agency. This contract is derived from GC/Works/1 Edition 3 1989 (revised 1990) and was intended to have substantially the same effect, but the provisions that would only be appropriate for central government employers have been suitably amended. For instance, the number of decisions that are final and conclusive was greatly reduced. It is not likely that PSA/1 will be revised to take account of the HGRCA.

Chapter 12
Engineering Forms of Contracts

ICE Conditions of Contract 6th Edition 1991

Generally

The dispute resolution mechanism found in the ICE 6th Edition, taken over from previous editions of the ICE Conditions of Contract, is for two-stage dispute resolution under clause 66. Before arbitration, an engineer's decision is necessary. A major change between the 5th and 6th Editions is that under clause 66(5) of the latter the ICE Conciliation Procedure (1994) is included. The purpose of the procedure is to provide an alternative to arbitration that is both quicker and more cost effective. The parties can, in the first instance, choose whether a dispute should be referred to a conciliator or should proceed to arbitration. If either party serves a notice to refer (i.e. to arbitration) the conciliation option is lost and the arbitration procedure has to be followed.

Settlement of disputes

The settlement of disputes is provided for in clause 66, which relates to all disputes arising from or in connection with the contract. However, although there are no exclusions stated in this clause, disputes over value added tax that is chargeable are excluded under clause 70(4). The timetable laid down in clause 66 needs to be followed carefully because certain choices may otherwise become unavailable or a recommendation of the conciliator becomes binding.

Under clause 66(2), a notice of dispute has to be served on the engineer. This should avoid uncertainty as to whether a dispute has arisen or whether a matter referred to the engineer has been referred under clause 66: *Monmouth County Council* v. *Costelloe & Kemple* (1965). It may still be alleged that the notice of dispute is premature, e.g. the engineer or the employer may contend that a claim has not been properly formulated so as to require a response. Alternatively, a party which has failed to give a notice to refer under clause 66(6) may seek to challenge the validity of the whole process by attacking the legitimacy of the notice of dispute.

Engineer's decision

After service of a notice of dispute the engineer has to give his decision in writing within one calendar month (clause 66(6)(a)(ii)) or within three calendar months if a certificate of substantial completion for the whole of the works has been issued (clause 66(6)(b)). The engineer's decision is final and binding, unless the subsequent recommendation of the conciliator has been accepted by both parties or the engineer's decision is revised by an arbitrator's award. Until such time, the parties are required to give immediate effect to an engineer's decision except where the contract has been determined or abandoned.

The date of receipt of the engineer's decision or the expiry of the period allowed for the giving of the decision triggers the next options available to the parties, if dissatisfied. These are:

(1) to request that the dispute be considered under the conciliation procedure (clause 66(5)), *or*
(2) to serve a notice to refer to start the arbitration procedure (clause 66(6)).

Conciliation can be sought where either party wishes to challenge the engineer's decision or where the engineer has failed to give a decision within the time allowed and so long as a notice to refer to arbitration has not been given.

Arbitration is possible in three situations:

(1) in the event of an engineer's decision under clause 66(3);
(2) if the engineer fails to give a decision;
(3) against the recommendation of a conciliator under clause 66(5).

The period in which matters can be referred to arbitration is ordinarily three calendar months following the decision, or on the expiry of the one-month period during which the engineer may make a decision. In the case of an appeal from a conciliator's recommendation, the period is one month, failing which the conciliator's recommendation becomes final and binding. Both employers and contractors should be mindful of this limitation.

Prior to the publication by the ICE of its *Corrigenda* (August 1993), there was a gap in the ICE 6th Edition whereby a contractor who had lost the right to arbitrate could invoke the conciliation procedure and, if dissatisfied with the conciliator's recommendation, still commence arbitration within one calendar month of receiving the recommendation. This loophole is now closed provided the *Corrigenda* are included in the contract.

ICE Conditions of Contract 6th Edition Revised 1998

Generally

The ICE Conditions of Contract Standing Joint Committee issued revisions on 19 March 1998 to incorporate amendments to cover the requirements of the HGCRA 1996. The clause 66 provisions of the 6th Edition 1991 have been completely revised in order to provide a contract that complies with the Act. The new clause 66 is entitled 'Avoidance and Settlement of Disputes'. The CCSJC recommends that the amendments are incorporated but if the project is not subject to the HGCRA the parties may choose not to incorporate these amendments and to use the existing provisions. If the parties did not incorporate these amendments but the project was subject to the HGCRA, the Scheme for Construction Contracts would become operative.

Disputes

The new provisions are provided both to overcome disputes and to define disputes to enable their early resolution. It is not possible to overcome the causes of all disputes but procedures can obviate certain problems and this is what is intended. Initially, the parties are given the right to seek an engineer's decision in respect of specified matters on which they are dissatisfied. The concept of the engineer's decision is maintained in this revision and the employer can seek such a decision where dissatisfied with any of a range of matters referred to in clause 66(2). The contractor has a similar right, which also extends to an act or instruction of those acting on behalf of the engineer. The engineer must provide a decision in writing within one month of the referral of any matter under clause 66(2).

During this initial stage of dissatisfaction there is said to be no dispute and subclause (3) spells this out. By definition a dispute cannot arise until one month has elapsed from the notice of dissatisfaction, or an earlier engineer's decision has been given and it is either unacceptable or has not been implemented and a notice of dispute has been served. A dispute also arises where an adjudicator has given a decision and where a notice of dispute is given because the decision was not implemented. To some, this may appear contrary to common sense: as soon as they are dissatisfied with a matter they may see it as a 'dispute'. There is significance about the wording on two counts. First, until a dispute has arisen there is no right to adjudication, and secondly, it is an attempt to diffuse matters by providing for a review which may dispose of the issue. An engineer's decision is therefore an essential prerequisite unless the time for such a decision to be given has passed. This does not in any way offend the legislative requirement of a right to adjudication because this only occurs once a dispute arises.

Unless the contract has been determined or abandoned, the parties must continue to perform their obligations regardless of the existence of a dispute. They must give effect to an engineer's decision under clause 66(2) and to an adjudicator's decision under clause 66(6), until it is changed by agreement or revised by arbitration. In this respect, agreement also includes the recommendation of a conciliator unless referred as discussed in the next paragraph.

Conciliation

Conciliation as a means of settling disputes has been retained (clause 66(5)) in a now rather cluttered disputes procedure, and its operation is broadly unchanged. The parties may agree for the dispute to be considered under the ICE Conciliation Procedure (1994) at any time prior to the issue of a notice to refer to arbitration under clause 66(9). Conciliation can only be commenced once a dispute has arisen and an engineer's decision obtained. A conciliator's recommendation becomes final unless it is challenged by issuing a notice of adjudication or a notice to refer to arbitration within one month of the recommendation.

Any matters brought before the conciliator are without prejudice. This promotes openness (or should do) while preventing subsequent exploitation, by one party calling the conciliator as a witness in subsequent proceedings. This is specifically prevented in clause 66(12)(b) and rightly so. Whether parties will keep their cards to their chests is not beyond doubt.

Adjudication

This revision contains in clause 66(6) a new right for either party to refer to adjudication a dispute under the contract. Adjudication is commenced by serving a notice in writing, referred to as the notice of adjudication, which brings into effect the ICE Adjudication Procedure (1997) discussed later in this chapter. A dispute must exist before adjudication may be started and where one does exist, it is possible in theory for conciliation and adjudication to be simultaneous.

The detailed adjudication provisions set out in clauses 66(6)(b–f), 66(7) and 66(8) are drafted to give effect to the requirements of section 108(2–4) of the HGCRA. The clauses mirror the wording of the Act – see Chapter nine for a detailed commentary.

The parties choose, in the first instance, whether a dispute should be referred to a conciliator or should proceed to either adjudication or arbitration. As in the 1991 edition, if either party serves a notice to refer (i.e. to arbitration) the conciliation option is lost and the arbitration procedure has to be followed, unless otherwise agreed. In addition, the parties have a right to adjudication which may be exercised at any time once a dispute has arisen, even if conciliation has been commenced or a notice to refer to arbitration has been given.

The engineer may be called as a witness in any dispute resolution proceedings regardless of an appearance at any earlier dispute forum under the contract.

Arbitration

All disputes, with the exception of the failure to give effect to an adjudicator's decision, can be finally determined by arbitration. The specific exclusion enables a party to seek enforcement of the adjudicator's decision in the courts. Arbitration is started by serving a notice to refer and the arbitration may be commenced at any time whether the works are complete or not. Where an adjudication has taken place and a decision has been given, such a notice must be given within three months. If such a notice is not given, the adjudicator's decision will be both final and binding on the parties. Therefore this time should be strictly adhered to if a party wishes to avoid this consequence. The period in which a conciliator's recommendation can be challenged in arbitration is shorter.

The other arbitration provisions follow the earlier edition except that reference is made to the arbitrator's ability to open up review and revise any decision, opinion, instruction, direction, certificate or valuation of an adjudicator. Furthermore, the arbitrator may hear evidence and arguments in addition to those put to an adjudicator.

ICE Design and Construct Conditions of Contract 1992

Under the ICE Design and Construct Conditions of Contract there is a similar mechanism to that in the ICE 6th edition but with two important differences. First, there is no engineer and therefore no reference to an engineer's decision. Secondly, the conciliation procedure is not optional but obligatory before arbitration can be commenced; that is, unless the parties subsequently agree otherwise.

A notice of dispute must be served in accordance with clause 66(2). Clause 66(3) then provides for conciliation, if a dispute is not resolved within one calendar month from service of the notice of dispute. The one-month period before the conciliation procedure becomes operative allows the parties to reconcile their differences by whatever means they can agree. After one month has elapsed, the dispute, according to clause 66(3), 'shall' be referred but in reality 'may' be referred to conciliation in accordance with the conciliation procedure; the conciliator has to make a recommendation in writing within three calendar months of the service of the notice of dispute.

The conciliator's recommendation is final and binding, unless the recommendation has been revised by an arbitrator's award in accordance with clause 66(5). Until such a time, the parties are required to give immediate effect to the recommendation except where the contract has been determined or abandoned.

Under Clause 66(5) arbitration is possible in two situations:

(1) Where either the employer or the contractor is dissatisfied with any recommendation of a conciliator appointed under subclause (3), *or*
(2) Where the conciliator has failed to give a recommendation for a period of three calendar months after service of the notice of dispute.

ICE Design and Construct Conditions of Contract Revised 1998

The ICE CCSJC issued revisions on 19 March 1998 to incorporate amendments to cover the requirements of the HGCRA 1996. These revisions are virtually identical to those for the principal ICE form and the discussion above of the ICE Conditions of Contract 6th Edition Revised 1998 applies to the design and build form. The only difference is the reference to the employer's representative instead of the engineer. The clause 66 provisions of the 1992 edition have been completely revised to provide a contract that meets the requirements of the HGCRA.

ICE Conditions of Contract for Minor Works 2nd Edition 1995

Clause 11 of the ICE Conditions of Contract for Minor Works provides a mechanism that is similar to that contained in the design and construct conditions of contract, although here there is an engineer. The engineer's decision is discarded.

Clause 11.2 provides for a notice of dispute to be served, stating the nature of the dispute. Either party may then within 28 days of the service of the notice give written notice requiring the dispute to be considered under the conciliation procedure. So long as the dispute has not already been referred to a conciliator under clause 11.3, the parties can have the dispute referred to arbitration under clause 11.5. This is done by service to the other party within 28 days of the notice of dispute of a further notice, the notice to refer. The notice to refer is an indication that the aggrieved party wishes the matter to be resolved by arbitration.

Unlike the Design and Construct Conditions of Contract, there is no waiting period before either conciliation or arbitration can be commenced. It is anticipated that at the time of giving the notice of dispute a notice will be also given under either clause 11.4 or 11.5. Furthermore, conciliation is optional and not obligatory as in the design and construct conditions of contract.

If the notice to refer is not served within 28 days of the notice of dispute and no notice in writing is given requiring conciliation, then the notice of dispute is deemed to have been withdrawn. Nevertheless, it is still possible to issue a further notice until 28 days after the issue of the final certificate. Clause 7.7 provides for a notice of dispute, except on VAT matters, on the final certificate to be given within 28 days of its issue.

Regard should also be paid to Clause 11.4. If the parties decide to adopt the conciliation procedure either party may, within 28 days of the receipt of the conciliator's recommendation, refer the dispute to arbitration by service of a written notice to refer. If the notice to refer is not served within 28 days, the recommendation of the conciliator is deemed final and binding on the parties. A reference to arbitration is to the Institution of Civil Engineers' arbitration procedure.

ICE Conditions of Contract for Minor Works 2nd Edition 1995 Revised 1998

The ICE CCSJC issued revisions on 19 March 1998 to incorporate amendments to cover the requirements of the HGCRA 1996. These revisions are virtually identical to those for the main ICE form and the discussion above of the ICE Conditions of Contract 6th Edition Revised 1998 equally applies to the minor works form. Clause 11.1 of the 1995 edition has been revised to make reference to addendum A, which now contains the dispute resolution procedures. The reference to clause 7.9 has changed to 7.11 following a simple renumbering of clauses. The remaining clauses 11.2 to 11.7 are deleted and addendum A is substituted in order to provide a contract that meets the requirements of the HGCRA.

Addendum A is virtually identical to the new clause 66 of the ICE main form but each clause is prefixed with A.

ICE Conciliation Procedure 1994

Turning to the mechanics of the conciliation procedure, this is set out in three sections. First, a general preface explains the main differences between conciliation and arbitration; secondly, the conciliation rules are set out in 23 numbered paragraphs; and the third part of the conciliation procedure booklet is a conciliator's agreement together with a schedule in which details of fees can be set out. The rules seek to achieve maximum clarity with the minimum of technical legal language. In this aim they are largely successful.

The aims of the panel when drafting the conciliation procedure (1994) were to emphasise:

- achieving a settlement;
- the non-binding nature of the process;
- the confidential nature of the process;
- the process is 'without prejudice' and anything said or conceded will not affect the parties' positions if a settlement is not reached;
- the 'recommendation' is not intended to be a judgement but an opinion as to how the matter should be settled in the best interests of the parties.

The conciliation procedure can be used in two situations. First, it obviously applies where particular conditions of contract expressly state it to apply. Secondly, the parties may choose, although not an original contractual obligation, to adopt the conciliation procedure later. The spirit of the procedure is expressed in rule 2 as follows:

'This Procedure shall be interpreted and applied in the manner most conducive to the efficient conduct of the proceedings with the primary objective of achieving a settlement of the dispute by agreement between the Parties as quickly as possible.'

The object of the procedure is an expedited result. Rule 4 states that the parties either will agree the appointment of a conciliator within 14 days or, in the event of a presidential appointment, the appointment should occur within 14 days of a request first being made. Under rule 6, the party requesting the conciliation undertakes to provide to the conciliator immediately following his appointment, and simultaneously to the other party, any copy notice of conciliation together with all relevant copies of notices of dispute and other documents which are conditions precedent to conciliation. The instigating party may provide the conciliator (rule 8) with a statement of its views on the dispute and the issues which it wishes to see considered by the conciliator. It is then the conciliator's responsibility to arrange a conciliation meeting with the parties (rule 9). The parties must provide to the conciliator seven days before the conciliation meeting, with copies to the other party, details of the persons who will attend the conciliation meeting together with an indication that those persons have authority to act on behalf of the party (i.e. make binding decisions).

Rule 10 aims to achieve maximum flexibility in the conduct of the conciliation by permitting the conciliator to take, with the consent of the parties, legal or technical advice. The rule also permits him to investigate the facts and the circumstances of the dispute. The conciliator is permitted, under rule 11, to have independent discussions with either or both of the parties, although confidential information obtained during the course of such meetings may only be released to the other party with the consent of the party making the disclosure to the conciliator.

Under rule 12, a party may ask that additional claims or disputes or additional parties shall be joined in the conciliation. This is not carte blanche to bring in any other matter because it is necessary to provide all appropriate information and most importantly the notice of dispute and decisions. It would be unwise for the conciliator to consider additional items without ensuring there has been compliance with the mechanism provided in the contract.

Where the process of conciliation would be aided by further investigations, the conciliator may adjourn proceedings (rule 13).

Rules 14–17 deal with the question of reaching agreement. Under rule 14 the conciliator may assist the parties in the preparation of an agreement setting out the terms of settlement, while rule 15 recognises the possibility that the parties may be so polarised that settlement cannot be achieved. Rule 15 of the procedure deviates from non-binding forms of ADR. The conciliator is empowered, in the absence of agreement between the parties, to issue a recommendation. The nature of a recommendation is set out in rule 16. This is the conciliator's solution to the dispute, based on the conciliator's opinion as to how the parties can best dispose of the dispute between them. Unlike an arbitrator or judge, the conciliator can be practical and is not bound to adopt strict principles of law or give reasons for his recommendation, although the conciliator may within seven days of making his recommendation give reasons for it.

Rule 17 deals with the question of the conciliator's fees and disbursements which, unless there has been a separate agreement between the parties, should be a joint and several responsibility of the parties to be paid within seven days of the receipt of the account. Receipt of payment means despatch by the conciliator to the parties of his recommendation. This is broadly similar to the release of an arbitrator's award.

The conciliator may review and revise his recommendation, subject to payment of additional fees by the parties. An important provision is found in rule 19, which permits the conciliator to participate in subsequent arbitration proceedings as arbitrator if the parties agree. This allows the conciliator to become involved in a form of Med-Arb, discussed in Chapter 7. Importantly, and quite properly, the conciliator cannot be a witness for either of the parties in any subsequent litigation or arbitration. The one remaining provision of some note in the conciliation procedure is rule 21 which overcomes the question of experts/adjudicators not enjoying, at least at common law, an equivalent immunity to that of judges and arbitrators. As experts or adjudicators (including conciliators) are potentially liable to those who employ them for negligence or breach of contract claims, rule 21 acts as an exclusion of liability but does not extend to cover third party actions.

ICE Adjudication Procedure 1997

The conciliation and adjudication advisory panel of the Institution of Civil Engineers produced this procedure in response to the HGCRA, for use in connection with the ICE forms of contract and to be administered through the dispute administration service. Although the adjudication procedure was developed for ICE forms of contract, a preamble states that it may be suitable for use with other contracts. The procedure comprises nine clauses plus a form of adjudicator's agreement and an attached schedule, together with a standard notice of intention to refer a dispute and a standard form of application for the selection and appointment of an adjudicator. A published list of adjudicators is also available that usefully provides brief curricula vitae and the areas of dispute in which they operate.

Clause 1 sets out the general principles under which the procedure is to be conducted. The object is to achieve a fair, speedy and inexpensive resolution of the dispute. An adjudicator must be a named person; the appointment of a firm is considered inappropriate. An adjudicator may rely on his own expertise and may be 'proactive' in ascertaining the facts to arrive at a decision. Where a conflict emerges between the contract under which the dispute has arisen and the adjudication procedure, the latter will take precedence.

An adjudicator appointed under the procedure, together with any employee or agent, is immune from liability for his acts or omissions and the parties provide an indemnity against third party claims. The procedure also excludes the ICE's liability in respect of the appointment of an adjudicator or conduct of the adjudication.

The adjudicator cannot be subsequently appointed as arbitrator unless the parties agree. In any event, it is unlikely that an adjudicator would be appointed arbitrator unless chosen by an appointing body in ignorance of the adjudication. It is unlikely that parties would agree to the adjudicator acting as arbitrator because one party will already have received an adverse decision, albeit perhaps on a different dispute.

Because an adjudicator is privy to much information which under the procedure is confidential, the parties cannot call the adjudicator as a witness in any legal proceedings or arbitration concerned with the same subject matter. Interestingly, it does not preclude an adjudicator from a subsequent adjudication where another adjudicator has been appointed.

Notice of adjudication

When a dispute arises under the contract, it may be referred to adjudication by issuing a notice of intention to refer a dispute. This is known as the notice of adjudication and a standard pro forma contained within the procedure can be used for this purpose. This form requires the insertion of basic information, names, addresses and the contract under which the dispute has arisen. In addition, the issues(s) in dispute are set out in Appendix A, and in Appendix B the remedy sought. The notice of adjudication recognises that the adjudicator may have already been named in the contract or, alternatively, that the referring party needs to name an adjudicator or adjudicators for consideration by the other party. The party issuing the notice selects the relevant option.

Where an adjudicator has already been named or agreed, a copy of the notice of adjudication should be sent to the adjudicator. The notice also requires the adjudicator to confirm a willingness to act within four days. Failing this, the ICE will be requested to make a selection/appointment. This tight schedule is to ensure that the procedure complies with section 108(2)(b) of the HGCRA. The selection and appointment must be dealt with together because the selection alone would not conform to the needs of this provision.

If the ICE is requested to make an appointment the standard form attached to the procedure may be used and completed by the party requiring the adjudication. This form, when completed, is sent with the notice of adjudication and appropriate fee to the dispute administration service. The only issue on this form requiring consideration is the choice of type of adjudicator with particular expertise. The form provides for the party to select from those with experience in civil engineering, building or process engineering. The ICE returns the form to the initiating party, identifying the selected adjudicator.

If the adjudicator has not been named or subsequently agreed on, the notice of adjudication will include the name or names of adjudicators for consideration by the other party. On receiving it, and within four days of the issue of the notice, the other party, where a choice is provided, must select or otherwise confirm acceptance and notify both the referring party and adjudicator. The timescale is extremely tight, aggravated because it runs from the date of issue of the adjudication and not its receipt.

Where there is a failure to secure appointment of an adjudicator under the provisions of clauses 3.1 and 3.2 within four days, the parties have a further three days during which they can implement the contract provisions to appoint an adjudicator. Where no appointing body is named the procedure provides for the ICE to make the necessary appointment at the request of either party.

Where an adjudicator has been appointed but subsequently becomes unable to act, either party may ask for a replacement to be nominated, either under the terms of the contract or by the ICE. This begs the question of whether the original adjudicator would have any liability for not acting under the agreement. It might depend on the reason.

Adjudicator's agreement

An adjudicator should always execute the adjudicator's standard agreement and schedule before carrying out the adjudication. This is also a specific requirement of the procedure. The timing of this will depend on whether the adjudicator has been named in the contract but both parties are obliged to sign such an agreement within seven days of being requested to do so. This obligation would be enforceable as it is included in the ICE adjudication procedure which is itself incorporated by reference in the principal contract. The ICE adjudicator's agreement provides that the adjudication shall be conducted in accordance with the procedure and that the parties are jointly and severally liable to pay the adjudicator's fees and expenses. It also provides that the adjudicator and the parties shall maintain confidentiality. However, it is possible to make public certain information, if the consent of the other parties is secured. The adjudicator also has a duty towards any documents received for the adjudication and must inform the parties if he intends to destroy them. No period is stated during which the parties can ask for the documents to be retained, but it would be sound practice before destroying these documents for the adjudicator to receive written confirmation that there is no objection.

Attached to the adjudicator's agreement is a schedule which sets out the hourly rate, the definition of expenses and that VAT is chargeable, where appropriate. The schedule sets out when the adjudicator should be paid and that outstanding amounts will be subject to interest at 5% over the Bank of England Base Rate. Payment has to be made within seven days of the invoice and the decision cannot be made dependent on payment. However, under clause 6.6 the adjudicator can refuse to deliver the decision without full payment of the fee and expenses. The notice can only be given up to seven days before the decision is due, i.e. 21 days after referral. It would appear that this does not accord with the HGCRA which requires that the contract must provide for a decision within 28 days of referral or such longer time as is agreed or permitted.

Presumably the adjudicator will invoice the parties separately for their equal shares unless the adjudicator directs otherwise in the decision (clause 6.5). Where delivery of a decision is made conditional on payment, either party may pay the adjudicator and recover the other party's share of costs, if any, as a debt. The hourly charge applies not only to the time spent on the adjudication but also to any incidental time, including travelling time. Choosing a local adjudicator with expertise in the type of dispute could therefore appreciably affect costs.

The schedule also provides for an appointment fee, which is determined by the parties and payable in equal shares within 14 days of the appointment. This takes effect as an advance payment that may be refundable in whole or part. It is not a retainer fee as such. Nevertheless, once the adjudicator has entered into the agreement he is obliged to provide a decision in respect of disputes referred, regardless of whether an appointment fee has been stated and paid. It is of advantage to the parties to have selected an adjudicator at the early stages of a project. The adjudicator may be reluctant to enter an agreement well in advance of a dispute because this will create an obligation to act. Clause 3.1 recognises the reality of this and requires the named adjudicator to confirm a willingness to act within four days of receiving the notice of adjudication.

Adjudication

Once it is known that the adjudicator will act, the referring party has two days in which to send a full statement of the case. Clause 4.1 refers specifically to:

- the notice of adjudication
- adjudication provision within the contract
- information and documents to be relied on.

Although the timescale appears short, the referring party need not submit a dispute notice until ready, subject of course to the one-month time limit running from a conciliator's recommendation. By contrast, the responding party has only 14 days in which to make a response to the adjudicator unless an agreement to extend time can be agreed with the other party and adjudicator.

The receipt of documents by the adjudicator is important. This sets in motion the 28 days for the decision to be given. The adjudicator is obliged immediately to inform the parties of the date of their receipt. The time for arriving at a decision may be unilaterally extended by up to 14 days by the referring party or by any mutually agreed period so long as the agreement is made after the dispute is referred. Any such agreement made prior to the dispute would fall foul of section 108(2)(c) of the HGCRA.

The powers of the adjudicator are wide and are set out in clause 5. These include a total discretion as to the conduct of the adjudication except as otherwise provided. This means that strict rules of evidence need not be followed. There is power to open up and review all matters under the contract. However, an adjudicator cannot review a previous adjudicator's decision unless the parties agree. This restriction is imposed because any appeal against an adjudicator's decision should be in arbitration. Nevertheless, there is the possibility that parties might subsequently agree that an adjudicator could review a previous decision, although it would seem unlikely that this would occur.

The adjudicator can enlist the help of a legal or technical adviser but before doing so must advise the parties because they are responsible for the fees incurred. It is unclear what would happen if one or both of the parties were unhappy with this course of action. The procedure suggests the adjudicator could go ahead, having fulfilled the notification requirement, but the potential problems are obvious. Once appointed, the fees incurred by such advisers are payable if the parties have received the benefit of the advice. Parties to the adjudication bear their own costs and this may provide restraint on the employment of expensive personnel.

Decision

The adjudicator's decision can be given in parts if different issues are to be considered, so long as the overall time is honoured. Even if the decision is not given within the prescribed period, the decision shall still be effective so long as it is given before the dispute is referred to a replacement adjudicator. If an adjudicator fails to meet the time limit, either party may give seven days notice of his intention to seek a replacement adjudicator. Such a notice would almost certainly have an impact and a decision would be likely to be forthcoming unless there were some insuperable reason to the contrary. However, to appoint a replacement raises issue of judgment. Would the party seeking the replacement be best served by such an action? It could result in an even later decision or could go against the party who sought the replacement.

Generally, the earlier the decision the better and it should be given. There is no requirement on the adjudicator to provide reasons for any decision but the decision itself may award whatever is considered to be an appropriate remedy, including simple or compound interest. Interest is at the discretion of the adjudicator but any amounts considered outstanding and the timing of the referral in relation to the event causing the dispute, will be influencing factors. In the decision, the adjudicator may also direct that his fee and expenses are paid other than in equal shares. However, the default position is that the fees and expenses are paid in equal shares and for the adjudicator to deviate from this position would require a good reason. The adjudicator would need to be convinced that the adjudication was unnecessary in some way or that one party was acting unreasonably before directing that fees, etc. be apportioned differently.

An adjudicator's decision must be implemented immediately, regardless of whether it is to be subsequently challenged. It is binding until finally determined in legal proceedings, arbitration or by agreement. Where the decision involves payment, the adjudicator can direct how and when this is made unless it relates to a notice under section 111(4) of the HGCRA, otherwise it becomes payable in the next 'stage payment' (clause 1.6). The use of the term stage payment is presumably intended to include both monthly and stage payments, even though they may be distinguished.

The adjudicator or either party has the ability within 14 days of its notification to ask for the decision to be corrected. This is quite different from a review of a decision and is limited to clerical mistakes, errors and ambiguities (akin to the arbitration 'slip rule'). The adjudicator makes any corrections within seven days of the request but what the position is if he fails to do so is not stated. It is open to the parties to agree corrections.

FIDIC Conditions of Contract for Works of Civil Engineering and Construction Fourth Edition 1987

Generally

This contract is for construction works where tenders are sought on an international basis. The foreword to the contract states that it is also suitable for domestic contracts, subject to minor modifications. The use to which this contract is put may be important because of the applicability of the HGCRA to the UK (the effect of the HGCRA is extended to Northern Ireland by Order in Council in accordance with section 149). If the works are carried out in the UK, regardless of by whom, this Act would apply. FIDIC does not comply with the requirements of the HGCRA, so the Scheme for Construction Contracts would apply. The Act also applies to works even where the law of the contract, as stated in subclause 5.1 (see Part 11 of the form), is not that of England and Wales or Scotland. Conversely, the HGCRA does not apply where works are not carried out in the UK, even if the law inserted in subclause 5.1 of FIDIC, is that of England and Wales or Scotland. This is because the HGCRA does not affect work outside the UK. The location of the works is important.

Engineer's decision

Clause 67 provides for the settlement of disputes. In the first instance, where any dispute arises between the employer and contractor, it shall be referred to the engineer for a decision. Clause 67.1 embraces the whole range of disputes that may emerge, including those arising from the engineer's own instructions, certificates and valuation. It also covers those that arise post repudiation or after some other form of termination. To allow the engineer to make a decision that may have arisen because of his own action is seen by some as bizarre and an encouragement to cover things up, but it does provide an opportunity for reconsideration before more formal processes are adopted. The engineer's review is commenced by either party by giving written notice to the engineer and the other party.

Where a notice is given, the engineer has 84 days from the reference to give a decision and it must be stated that it is given under clause 67.1. The period for giving the decision runs from the day after the reference is made, therefore the 84 days are actually extended by one day. Such a period is long and by modern standards, where there is increasing desire to resolve disputes more speedily, it seems extravagant. During this period the parties must comply with the contract requirements as if no dispute had arisen and continue to do so until it is revised by the engineer, amicable settlement or by an arbitrator. The engineer's decision must state that it is given under clause 67.1 to avoid uncertainty as to whether such a decision has been given.

Where an engineer's decision has been given following a notice and either party is dissatisfied with the outcome, he may give notice of intention to commence arbitration. Such notice may also be given where the engineer fails to give a decision within the 84 days. If the engineer's decision is to be challenged it must be within 70 days after the day when the decision is received. Where the engineer has failed to give a decision the period runs from the expiry of the 84-day period. If no such notice is given within the stated periods, the engineer's decision becomes final and binding.

A valid notice challenging an engineer's decision is necessary before arbitration can be commenced, subject only to the application of subclause 67.4. Therefore, if a party is unhappy with a decision they must act accordingly, otherwise this right will be lost and the decision will hold. It is possible that neither party will challenge the engineer's decision but will then fail to comply with that decision. As the decision was not challenged, it would have become final and binding and therefore not usually appealable. However, clause 67.4 provides for this eventuality and despite no notice of arbitration and no attempt at amicable settlement, the parties may refer such a failure to arbitration; but as this is without prejudice to their other rights, a party may choose to seek summary enforcement through the courts.

Amicable settlement

Even where notice of arbitration has been given, proceedings cannot be commenced until the 56th day, or other agreed time, after giving notice unless the parties have attempted to settle their differences. Clause 67.2

imposes an obligation on the parties to try to resolve their differences amicably, but where they do not, arbitration cannot begin immediately. This moratorium gives the parties the opportunity to settle their differences amicably – a last chance saloon. If an effort is made to resolve their differences but the parties realise they will not come to a solution, they can agree to commence arbitration within the 56-day period. Interestingly, by strict definition the clause does not prevent a party unilaterally commencing arbitration if of the opinion that the parties have attempted to settle the differences amicably. If arbitration were commenced in such circumstances, the other party could resist by seeking to stay proceedings and it would be for the arbitrator to decide whether this was appropriate in the circumstances. The parties also have the right to agree to extend the 56-day period if they wish.

Arbitration

Ample opportunity exists under this form of contract to resolve disputes without redress to arbitration. The review of the engineer's decision and amicable settlement are the two mechanisms provided, but where these fail arbitration is available. The arbitration is conducted under the Rules of Conciliation and Arbitration of the International Chamber of Commerce, unless otherwise amended, as illustrated in Part 11 of the conditions. Subject to the procedures discussed above, arbitration can be commenced at any time. If arbitration ensues, the parties are not constrained by what has been put previously to the engineer and the engineer may be called in the arbitration proceedings.

Supplement to Fourth Edition of FIDIC First Edition 1996

This supplement was published in 1996 and included three sections: dispute adjudication board, payment on a lump sum basis and late certification. Section A deals with the dispute adjudication board and is the only one of interest in the context of this book. The supplement also contains useful guidance notes relating to disputes and arbitration, together with model terms of appointment and procedural rules for the dispute adjudication board.

Under FIDIC it is usual for the engineer to make decisions and to have the power to review decisions which are challenged by one or both parties. As suggested above, this is seen by some to be rather a bizarre situation and alternative solutions have been sought. Section A of the supplement recognised this development and provided the alternative of using a dispute adjudication board. An adjudication board can comprise one or three members and where it is thought desirable this alternative may be used and an amended clause 67 adopted.

Dispute review board

Where it is decided to replace the engineer as the decision-maker under clause 67 with an adjudication board, the clause must be amended. This will also entail an amendment to the appendix to tender. The decision on whether to use a dispute review board is difficult (see Chapter 6 for discussion) but will often turn on personal preferences and experiences rather than hard fact.

If a dispute review board is chosen, will it be a one person board or a three person board? This should be decided and inserted in the appendix. It will be determined by such factors as duration, size and complexity of the works, but there are no hard and fast rules. The complexity of the works may mean it is desirable for a number of persons with different technical qualifications to be available to consider the range of disputes that may emerge. There is nothing in principle to prevent the appointment of a number of one person boards to cover differing disputes on the same project. This is not a recommended route because it is fraught with difficulties, i.e. the classification of disputes that have arisen and delays in making a series of appointments. Too much concern over how many persons will form the board is unnecessary, because with the agreement of the parties the board can always seek other advice to fill in any gaps in knowledge or expertise. It would seem that a three person board is only necessary on large-scale complex projects.

The constitution of the dispute review board and the names of its members can be agreed and inserted into the contract. There is no specific place for this but it makes sense because it is desirable to have the board in place at the outset. If a one person dispute review board is appropriate, the employer may insert a name in the tender document for the agreement of the contractor. The name, together with biographical data, is included in the schedule. If the contractor does not wish to accept the proposed person the tender form entry is deleted accordingly and the appointment will be subsequently agreed. It is also possible for the employer to propose a list of names and to require the contractor to select one. Such a list may also be provided where a three person board is proposed, in an attempt to agree names before entering into the contract.

Where the dispute review board has not been agreed at the time of forming the contract, the parties are obliged within 28 days of the commencement date to appoint one. The commencement date is defined in clause 1.1 as the date on which the contractor receives the engineer's notice to commence, pursuant to clause 41. The parties have to agree the person for a one person board. Where a three person board is appointed the parties each nominate one member for the approval of the other party and also agree the chairperson. Agreement of the chairperson may be difficult, and according to the guidance notes it should not be a person of the same nationality as either party, unless both parties are of the same nationality. Nevertheless, FIDIC considers agreement of the parties is preferable to parties' nominated members agreeing a third member to act as chairperson.

When one party suggests names to the other party this may pose difficulties related to the perception of partiality, and other means will sometimes need to be adopted. However, whatever means are used, agreement is necessary unless FIDIC itself makes the nomination(s). Where agreement is not reached on who shall be the member(s), the appointing body, usually FIDIC, will, after consultation with the parties, make the appointment. The president of the Fédération Internationale des Ingénieurs-Conseils or a person appointed by the president will be the appointor unless it has been otherwise stated in the appendix. There seems little reason to choose another body. Once the appointing body has made its decision, it is final and conclusive.

The terms of appointment of the board are based on the FIDIC Model Terms of Appointment for a dispute review board. There are separate but similar terms for one member and three member boards and the terms stated in clause 67.1 are repeated in this agreement. Once a member has been appointed they cannot be replaced except by agreement of the parties or because they have terminated their own employment. In the normal course of events the board members would remain in place until the discharge of final payment and the return of the performance security as referred to in clause 60.7.

In circumstances where a board member ceases to be available, the parties may agree a suitable replacement and this may also be done in anticipation of such an event occurring. Where agreement cannot be reached, the appointing body would fill the gap.

The dispute review board works differently from a straightforward adjudication in that the board maintains contact with the works throughout its duration. The parties should consider using the board not only as a means of resolving disputes but also as a prevention. Further discussion of this appears later when dealing with the board's agreement and the dispute procedure.

Referring disputes to the board

Any disputes arising under the contract are referred to the dispute review board and, as such, this is not an optional part of the dispute resolution process as is often the case. The reference to the board must be in writing with a copy to the other party, and to avoid any possible confusion it must state that it is made under clause 67.1. It is important that the party referring the dispute sets out clearly the matter in dispute and the nature and extent of the remedy sought. Regardless of the reference, the parties must continue to comply with the contract. The responding party has no specific right under the contract to reply to this submission but they will usually wish to do so and if they do not the board is likely to ask if they wish to respond. This matter is also referred to later under the section on procedural rules.

The board has a wide range of powers and can require the parties to provide additional information it feels is necessary to arrive at a decision. The time for making a decision is the same as that where the engineer is the decision-maker. Clause 67.2 requires the board to provide a reasoned decision but whether this extends to reasons being given with the decision is a moot point. The guidance note says that reasons are an essential part of a decision and the spirit of the contract, perhaps, implies that they should be given. The model terms of agreement do not clarify this point, so it is important to determine what is expected of the board when entering into the agreement.

Challenging the board's decision

The procedure for challenging a decision of the board is similar to, but not identical to, that provided for challenging the decision of the engineer under the 1987 edition.

Where a decision is not given within the 84-day period or where either party is dissatisfied with the decision given, the party can give notice to this effect up to 28 days from the date of the decision or the last date on which it should have been given. This notice must state that it is given in accordance with clause 67.2, together with the reasons for dissatisfaction and a copy must be provided to the engineer for information. The supplement refers to a notice of dissatisfaction, whereas the unamended clause 67 refers to a notice of intention to commence arbitration, but the effect in either case is the same. Where no such valid notice is issued, the board's decision becomes final and binding on both parties.

If a valid notice of dissatisfaction has been given, the parties are then required to try and settle their differences amicably before commencement of an arbitration. Arbitration proceedings cannot be commenced without such a notice and without first attempting to settle differences between the parties. This offers a sensible second stage approach. However, once a decision is given by the board, the favoured party is unlikely to want to permit the formulation of an alternative solution and therefore may not be well disposed to seeking an amicable settlement, nor to agreeing to commence arbitration earlier than required under the contract. Although the contract endeavours to encourage an amicable settlement there is no obligation on the parties to settle, and failure, or indeed the failure to attempt such a settlement, does not prevent the commencement of arbitration proceedings. Proceedings cannot be commenced until 56 days after the notice of dissatisfaction unless the parties agree otherwise.

An amicable settlement can be sought using any of the ADR techniques; the contract is not prescriptive. Clearly, amicable settlement will only work if the disputants want to find a solution.

Arbitration is available for the settlement of any dispute concerning a decision of the dispute review board, where notice of dissatisfaction has been given and where an amicable settlement has not been reached. It is also available immediately where there is a failure to comply with a decision of the dispute review board which has become final and binding or where the appointment of the board has expired. The commencement of arbitration will require some formal action on the part of one or both parties and this is not set out in the provisions. There is no time limit specified for the commencement of arbitration. This would be governed by limitation or prescriptive periods determined by the applicable law.

It is important to complete the appendix to ensure that should a dispute arise the parties can commence arbitration speedily, addressing the substantive issues and not spend time agreeing the process to be adopted. If the rules of arbitration of the International Chamber of Commerce (ICC) are not to apply, then the name of the institution to administer the rules should be inserted in the appendix to the tender.

FIDIC model terms of appointment for a dispute review board

As noted earlier, clause 67.1(a) states that the model terms of agreement should be incorporated but it also allows amendments to these provisions. The intention is that the parties enter into agreement with the board member(s); this is a sensible approach. Of the various agreements discussed in this book these model terms are the most comprehensive yet lack absolute certainty on one essential component: the requirement that members must act in accordance with the 'Procedural Rules'. The contract does not refer specifically to the 'Procedural Rules' but to the 'Model Terms'. The latter include a prohibition clause in item 3(e) that refers to the 'Procedural Rules' but this provision only refers to providing advice and the board is competent to do far more. Consequently, a direct reference to the incorporation of the 'Procedural Rules' and the board's obligation to follow them would be preferable.

The terms of appointment set out in clause 67.1 and the model terms require the board members to remain independent of the parties and to act impartially in accordance with the contract. Board members are under a continuing duty to give notice of anything which might be seen to compromise their position. In this regard, clause 3 of the terms, without prejudice to the general requirement of independence, sets out a list of circumstances that are considered to breach the requirement of independence and impartiality. Some circumstances can be overcome by providing information to the parties or by seeking their consent. Non-disclosure of some facts is considered wrong, even though they may not affect the adjudicator's decision. The listing of these key items is good practice and could be usefully adopted by others when drafting such agreements. The board can only give advice to the parties or the engineer in accordance with the procedural rules prepared by FIDIC. Furthermore, a board member is prevented not only from making agreements with the parties or engineer but also from having discussions concerning employment by them in any capacity. The board's independence is fundamental and to reinforce this, members are appointed in their personal capacities and not that of their firm. Moreover, they will not be entitled to any fees and expenses following a breach of these provisions.

A further important provision covers immunity of members of the board who shall not be liable for 'any claims for anything done or omitted in the discharge of such . . . functions unless the act or omission is shown to have been in bad faith', the parties indemnify the members accordingly. However, the board receives no protection from an action for misrepresentation in breach of the warranty given under clause 4, regarding their experience and other attributes.

Under clause 4(a) the member also agrees a potentially onerous obligation and one which would make some adjudicators unhappy, namely to ensure availability for all site visits and hearings. However, there is a questionable attempt to qualify availability by requiring 28 days notice (see clause 6(a)(i)).

Clause 6 sets out comprehensive provisions concerning payment of board members. A striking difference between this and many other forms of agreement is the use of a retainer fee but this is compatible with the requirement of being available throughout the project. The agreement states that the retainer fee is monthly and what the board member is expected to provide for the fee. The retainer fee is reduced by one half from the month following the 'Taking-Over Certificate' and ceases to be payable in the month following the end of the defects liability period. In addition to the retainer fee, a daily fee is paid for days spent working on the project, including travelling time, but excluding those items covered by the retainer. Expenses in connection with these duties are also recoverable. There is some potential overlap with items included in the retainer, but this is minimal and should not usually cause a problem. The amounts of the retainer fee and daily fee are variable and are to be determined by the parties. Where they fail to agree, clause 67.1 of the supplemental clause provides that the daily fee will be that payable for arbitrators under the International Centre for Settlement of Investment Disputes and the monthly retainer fee will be three times this amount. The use of a retainer fee suggests that the dispute review board is designed principally for large-scale projects.

The submission of invoices for payment is also covered in clause 6, with different mechanisms for the retainer fee and the daily fee. The former is invoiced quarterly in advance, whereas the latter is invoiced following a site visit or hearing. Although the payments are made equally by the parties, the contractor is required to pay the invoiced amount within 56 days of receipt and to recover one half from the employer through the normal certification process. If either party fails to make such payments the relevant default procedures in clauses 63 or 69 of the main form of contract can be implemented. This provides a very powerful means of securing enforcement and threat of its use would usually be persuasive. Alternatively, one party can maintain the board by paying all costs and recover one half of the costs from the other party. The alternative approach may seem more appropriate but if further costs are to be incurred there must be a good chance of securing recovery. Board members may suspend their services or terminate their appointment on non-payment of invoices.

A board member's appointment may be terminated where the parties agree and give reasonable notice. What constitutes reasonable notice is unclear and for the sake of clarity it would have been better to specify a minimum period. The termination would not in any way alter the rights of parties accrued from the board.

Members of dispute review boards are not usually employed under the contract as arbitrators and cannot do so unless the parties agree. Furthermore, such members cannot be called as witnesses in arbitration, unless the parties agree.

Finally, this agreement provides for resolution of disputes by a sole arbitrator, under the Rules of Conciliation and Arbitration of the International Chamber of Commerce.

Procedural rules for a dispute review board

It is clear from the way these rules are appended to the supplement, the reference in the model terms and the reference in clause 5 of the procedure itself that the intention is for them to apply to all disputes that come before the board. However, as discussed above, the drafting for the incorporation of these rules is not well done.

Items 1–4 of the procedure deal with the arrangements for regular site visits and for the provision of documentation. These show clearly that the dispute review board has a role beyond that of dispute resolution.

The procedure for settling disputes referred under clause 67 is set out in clause 5 and as the board members do not act as arbitrators they have a wide discretion for the conduct of a hearing and the nature of documentation to be provided. These rules provide each party with a reasonable opportunity of presenting their case but the respondent is not given an express right of response under the contract. Nevertheless, any board is unlikely to act otherwise as it would run in the face of fairness.

Once a dispute has been heard, the board formulates its decision in private and, wherever possible, provides a unanimous decision. Where this is not possible a majority decision can be given and the minority view expressed. Such a decision is more likely to be challenged than one that is unanimous. The board must give notice of its decision in accordance with clause 67.2 unless otherwise agreed by the parties.

The procedure also requires that the language to be adopted shall be inserted. This may appear a small point but on international contracts where a number of languages may be used it is essential.

The procedure for three person boards also covers the situation where two members may proceed in the absence of the third, unless the chairman gives specific direction not to proceed. The parties can, of course, overrule this direction if they wish but their agreement should be recorded.

IChemE Model Form of Conditions of Contract for Process Plant, Lump Sum Contract, 1995

Generally

This contract was first published by the Institution of Chemical Engineers in 1968 and revised in 1981 and 1995. The IChemE Form for Lump Sum Contracts has become known as the 'Red Book', and was specifically drafted to deal with the complexities of a project for a new process plant. However, the IChemE claims that the provisions are widely applicable for lump sum contracts in other process-related industries. The IChemE has produced a small suite of contracts and in addition to the 'Red Book' there is a contract for reimbursable contracts, known as the 'Green Book'; and a subcontract, known as the 'Yellow Contract', which contains provisions for use with either the 'Red' or 'Green Books'. Late in 1998 a minor works form, known as the 'Orange Book', was published. Although, these contracts are not 'mainstream' construction contracts, they are analogous.

The 1981 edition of the 'Red Book' made provision for disputes between the parties to be referred to arbitration but subject to the provisions for reference to an expert. The 1995 edition introduced some important changes to the dispute resolution provisions, with the principle of using an expert maintained. Clause 46 made provision for the settlement of disputes but this clause was subject to the provisions of clause 45 which enabled reference to an expert.

Guidance note EE provides some brief background on the use of arbitration and litigation and emphasises the desirability of having a quick and informal way of achieving fair decisions on relatively minor disagreements. It is for the latter purpose that clause 45 provides for an expert.

The Red Book and the Housing Grants, Construction and Regeneration Act

The 1995 contract was published prior to the HGCRA and obviously does not take account of the adjudication provisions of that Act. The IChemE form is designed to be applicable for work related to the supply and installation of plant and, as such, many of the contracts for which it will be used are exempt under the Act. Similarly, the Act does not apply to work outside the UK and although the form was primarily drafted for work in the UK, it is frequently used in other countries. In these circumstances, the fact that adjudication is not provided for would be irrelevant as the Act and the Scheme do not apply and would have no effect.

If a project falls within the ambit of the HGCRA, the IChemE suggests that an additional clause is inserted. This clause adopts the same wording as that contained within the more recently published 'Yellow Book':

'If the adjudication provisions of the Housing Grants, Construction and Regeneration Act 1996 apply to any dispute under the Contract, such provisions shall apply in addition to Clauses 45 and 46.'

The inclusion of this clause acknowledges that the Scheme for Construction Contracts would apply and that the parties have a statutory right to adjudicate. Should the parties wish to exercise this right, then adjudication will take precedence over the other dispute resolution provisions in the contract. The situation would not differ without such a clause but it highlights the significance of the Act and clearly communicates what is to happen in the event of a dispute. If parties wish to avoid the Scheme, appropriate amendments would be required which should refer to a published set of adjudication rules. The reference to detailed adjudication provisions could be covered by the inclusion of a special condition. The IChemE is in the process of publishing its own adjudication rules. The IChemE approach is novel, providing for the Act but retaining expert determination. Consequently, were a party to wish to proceed to adjudication it must take precedence over expert determination, although in practice both processes could run concurrently. Could a party use the findings of an expert determination in the adjudication? Probably yes. Interestingly, because of the HGCRA, expert determination could become optional for contracts covered by the Act.

On those contracts to which the HGCRA applies, the requirements of the Act may lessen the impact of expert determination as currently provided. In order to ensure expert determination remains significant, the contract would require some redrafting but even then, this may simply introduce a further layer of dispute resolution. The IChemE propose publishing a separate Model Form of Civil Engineering Subcontract for contracts that fall within the remit of the Act.

The use of an expert raises a number of potential legal problems[1]. The advent of the HGCRA and the role of the adjudicator may bring about a variety of subsequent changes in terms of the definition of an expert, an adjudicator and an arbitrator.

Reference to an expert

Bearing the above points in mind, this section considers the role of the expert under this form of contract.

Clause 45 makes specific provision for the appointment of an expert who is to be agreed between the parties. No provision is made for naming the expert in the contract but there is nothing to prevent the parties from naming the person in the special conditions or by amendment to clause 45. If this approach is not followed and the parties fail to agree, the president of the Institution of Chemical Engineers will make a nomination, if either party requests. The desirability of naming the expert prior to entering the contract is not obvious when the expert has potentially a number of different roles to perform. If the person is to be an expert, then his expertise must appertain to the nature of the dispute. If the potential disputes are varied then a number of experts may be required. The question would then arise as to whether it is appropriate to agree on more than one expert at the start and before signing the contract.

Although the guidance note states that the use of an expert is a quick and informal way of achieving fair decisions on relatively minor disagreements, the contract permits far larger problems to be referred. The disagreements that can be referred to the expert are those expressly provided for in the contract and any which the parties subsequently agree to refer. The express references under the contract embrace a range of significant matters, many of which are frequently contentious and require different kinds of expertise. However, although most engineers experienced in running projects of this type would probably argue that they have the range of expertise to perform each role identified for the expert, this may not be true of project managers. The 1981 edition referred to the engineer whereas the 1995 edition refers to the project manager. It will be interesting to see whether this alteration will affect the use of the expert determination procedures. The issues referable to an expert are wide and the express references in the contract are as follows:

> *Clause 14.2* – 'If either party does not agree with such extension then the matter shall be referred to an Expert...'
> *Clause 16.6* – covers disputes concerning the contractor's objection to a variation that is not accepted by the project manager.
> *Clause 16.7* – 'In the event that a dispute or disagreement arises between the Project Manager and the Contractor about any addition to or deduction from the Contract Price pursuant to this clause or clauses 17 (contractor's variations) or 18 (valuation of variations) or any modification to the obligations of the Contractor ... any such dispute or disagreement is not settled between them within a period of fourteen days, or such longer period as may be agreed, it shall be referred by them to an Expert...'

The two preceding clauses cover disagreements concerning variations, whether ordered or simply proposed. In addition, they cover disagreements on the consequential costs associated with the incorporation of the variation into the works and in respect of complying with the project manager's instructions concerning the preparation of proposals and a quotation for a variation.

The project manager's ability to order variations may be challenged by the contractor but this objection must be made in writing as soon as reasonably practicable and in the case of a variation order within 24 days of receipt. If this is not done the project manager has no obligation to act and the expert provisions cannot operate unless otherwise agreed. This would not of course affect the right to adjudication.

Under clause 17.1, the contractor has the ability to submit proposals for varying the works. The contractor, therefore, can be proactive in bringing about a variation but the project manager has absolute discretion over whether to accept such proposals, and any relevant decision cannot be challenged. The decision cannot be referred to either expert determination or arbitration. But whether it can be raised in adjudication is another matter and although technically it can, it would seem that the adjudicator may be reluctant to change the project manager's decision.

However, under clause 17.2 there is an important exception to this restriction and where the contractor states in his proposals that the variation is required to eliminate a specified hazard, this may be referred to the expert if the project manager decides not to issue an order or proposes an alternative solution. The clause reads:

> *Clause 17.2* – '...If the Contractor shall dispute the Project Manager's decision, the dispute shall be referred to an Expert...'
> *Clause 20.4* – 'If the Contractor disputes the disapproval by the Project Manager of any Drawing, such dispute unless resolved between the Project Manager and Contractor within a period of fourteen days, or such longer period as may be agreed..., shall be referred to an Expert...'

The contractor is required to issue drawings for the approval of the project manager who may disapprove of any drawing on the grounds that it does not comply with the express terms of the contract or that it is contrary to good engineering practice. Clause 20.4 gives the contractor the right to challenge the project manager's decision and to seek an expert determination.

> *Clause 33.7* – 'any dispute as to the issue or withholding of a certificate ... shall be referred to an Expert'

and a similar provision in respect of whether or not a final certificate should have been issued is covered by clause 38.5.

Clause 35.10 provides for a reduction to be agreed in respect of plant that does not comply with specified performance criteria but where agreement is not reached it will be determined by an expert.

Where the purchaser makes good defects that are the responsibility of the contractor a reasonable reimbursement is due and dispute concerning the amount shall be referred under clause 36.7 to an expert.

Disputes concerning suspension orders (clause 42.5), and costs at termination (clause 43.9) are also referable to an expert.

Status of expert and expert determination proceedings

Wherever the expert determination provisions operate, subclause 45.2 makes clear that the expert is not acting as an arbitrator. This means that the expert acting under these provisions does not share an arbitrator's immunity.

The expert may revise or overrule any decision or instruction of the project manager and both the purchaser and contractor are bound by the expert's decision. This decision and any directions that the expert may give are conclusive and their correctness cannot be questioned in any proceedings. Furthermore, once a dispute has been referred to an expert it shall cease to be referable for resolution under clause 46 and hence mediation is not available nor can the dispute be referred to arbitration. The dispute cannot be referred to arbitration nor can the decision of the expert be challenged unless set aside on grounds of misconduct. Therefore, the status both of the reference itself and the expert, because of the binding nature of the decision, were prior to the HGCRA, incredibly high. However, the process of referring the dispute to an expert is now in effect optional, in that, if the decision is not liked, a party may exercise their right to adjudication in an attempt to secure a different outcome. This would appear to be the case notwithstanding clause 45.7 because adjudication is not and cannot be covered in this clause.

The expert must determine disputes referred and the parties must continue to comply with his obligations under the contract because there is no right to suspend operations pending a decision. In addition, the parties are obliged to assist the expert by making available information requested and by facilitating access to the site and other premises. The expert's powers are wide and the decision can determine contractual issues, factual disputes and establish the amount to be paid.

In addition to the expert's power to determine the matters in dispute, there is the additional power to determine the fee payable for acting as expert. There is no power to determine how this will be borne by the parties. Clause 45.5 requires the expert's fees to be borne in equal shares by the parties. This differs from the earlier edition.

Although the expert is not governed by any additional rules of procedure under this contract, the rules of expert determination referred to in the Yellow Book could be adopted if the contract were suitably amended. This can be done by the parties entering into a form of agreement to accept the appointment of the expert, incorporating a requirement that the appointment and conduct of the expert in connection with the dispute shall accord with the IChemE's procedures and rules for appointment of experts.

Termination of the contract under clause 43 does not affect the provisions for reference of disputes to an expert or arbitration, whether or not a dispute has been referred prior to the termination.

Disputes

Clause 46 provides for the resolution of any claim, dispute or difference that may arise, subject to any claims that the contract makes referable to an expert or the parties agree should be so referred. Therefore, the parties, in many instances, could exclude the operation of this provision by referring disputes to an expert, but in practice disputes will have to be settled that are not considered suitable for resolution by an expert.

Where a dispute is referred under clause 46, the first level of resolution involves the parties themselves being obliged to 'attempt to negotiate a settlement in good faith', a non-binding obligation in law. It would be a good thing if this could be achieved but it does seem that it is not likely to succeed without some informal intervention. Nevertheless, the fact that both parties are obligated to try to settle their differences may help. If the issue cannot be resolved, the next stage is mediation. Subclause 46.2 provides that the parties can agree to refer the dispute to the Centre for Dispute Resolution (CEDR) or some other body. No time is stated in which the parties are expected to resolve their differences before proceeding to mediation and as agreement to mediate is an essential prerequisite this could be a long time.

Once a dispute has been referred to mediation, the parties have 28 days in which to reach a settlement before it is referred to arbitration. In circumstances where the parties cannot agree to proceed to mediation, they may go straight to arbitration.

Despite referring a dispute under clause 46, the parties must continue to comply with their obligations under the contract. There is no right to suspend operations pending a decision. Clause 46.7 provides further evidence that the completion of the project is the main objective and where possible, disputes will be resolved without disrupting progress to the works.

Arbitration

Arbitration should only arise where a dispute is not referred to the expert under clause 45. As a consequence, the arbitration provisions will be utilised far less than might otherwise have been the case.

Any disputes which are expressly provided as referable, or are agreed to be referable, under clause 45 cannot be submitted to arbitration. Any other disputes which may arise between the parties and are not settled by agreement or mediation must be referred to the decision of a single arbitrator. How the reference to arbitration will be conducted is not covered. The parties need to agree the arbitrator and failing this, the president or past president of the IChemE will make an appointment. In the 1981 edition there were 31 days to agree the arbitrator, but no time is inserted in this 1995 edition. It would seem sensible that some period is included, otherwise it may give rise to delay.

IChemE Yellow Book Subcontract 2nd Edition 1997

The second edition of this contract is substantially different from the first. The provisions in the second edition for reference to an expert, to mediation and arbitration are the same as the provisions in the 1995 Red Book, except for a few subcontract specific conditions, some updating, and the introduction of clause 47 dealing with the HGCRA. This clause is written in the contract and does not need to be inserted as for the Red Book.

Clause 45.1 is similar to the first edition but there is specific provision for the contractor and subcontractor to agree which other disputes may be referred to the expert. This clause provides that the expert determination shall be conducted in accordance with the rules of expert determination published by the IChemE, unless agreed otherwise before the appointment of the expert. The first edition of the rules of expert determination was published in 1998.

Once a dispute has been referred to an expert, it shall cease to be referable to arbitration under clause 46. This is different wording to that in the Red Book and although it still excludes arbitration, it does not prevent a subsequent mediation. It is unlikely that this is intentional because mediation requires agreement and therefore why should one wish to agree to mediation following a fully binding decision? The only circumstance where this may occur is where both parties are unhappy with the expert's decision.

Clause 45.9 permits joinder of the subcontractor to a dispute which is substantially the same as that arising between the contractor and the purchaser and the subcontractor agrees to such action.

Clause 46 contains similar provisions to the Red Book but subclause 46.3 states that arbitration shall be in accordance with the arbitration rules published by the IChemE, unless agreed otherwise, prior to the appointment of the arbitrator.

The new clause 47 establishes the relationship between statutory adjudication under the HGCR Act and the dispute resolution procedures in the contract.

IChemE Model Form of Conditions of Contract for Process Plant, Reimbursable Contracts, 1992

This particular contract is only briefly mentioned, as it was produced before the other contacts referred to above and is inconsistent in its provisions. Broadly, it is an updated version of the 1981 Red Book in that it provides for expert determination and arbitration but with a few adjustments to wording. There is no reference to mediation, nor is there any reference to statutory adjudication. However, as suggested by the IChemE and referred to above, a clause can be inserted.

IChemE expert procedures

In February 1998, the IChemE published a set of expert procedures for use with the model forms of conditions of contract for process plant[2]. This booklet is to assist parties and the expert appointed to understand the use of expert determination. The expert procedures comprise the 'Procedures and Rules for Appointment of Experts' which are in four parts, together with guide notes and annexes A–F. The four parts of the procedures are briefly described below.

Initial procedure to be followed

This provides the procedure for issuing the initial notice to appoint an expert, the agreement on the appointment, acceptance and confirmation by the expert.

Procedure to be followed in the event that the parties in dispute cannot themselves agree on the identity of an expert to be appointed

The rules set out that where the parties cannot agree on the expert, the party wanting the expert must write to the president of the IChemE who will make a nomination within 28 days of the application.

Rules to be followed when the expert has been appointed by agreement or by the President of the Institution

This section deals with the procedure on appointment and covers housekeeping matters as well as several substantive issues. The expert is required to write to each party directing them to provide both a summary and the detail, together with documentary evidence of the party's view of the dispute. The parties will also be asked to identify anyone who will be presenting evidence to support their view and anyone who will be representing the party at the preliminary meeting. No set time is stated for providing this information; the rule states that it shall be given in a reasonable time.

Other rules address matters such as the provision of copies of documents and are designed to ensure that the expert and the parties are in possession of all the documents to be relied on. Parties in dispute are not prevented from resolving their own problems but any such attempts must be reported to the expert except where a possible settlement is involved. These rules are detailed and the parties must provide a written account of discussions and any conclusions they may have reached. If settlement is being discussed the rules require that this is only reported when a settlement has been achieved. In these circumstances, the process of expert determination will stop but it is unclear what the expert's role would or should be, if the same dispute resurfaces because the 'settlement' reached was not honoured.

Under the rules, the procedure for investigation is at the discretion of the expert. The determination can take a variety of forms. The only requirements of the expert are to ensure there is no breach of the rules of natural justice and to give notice as to the procedure. The parties may be required to attend meetings and reasonable notice of these is required.

Powers of the expert

Rule 4A.1 states:

> 'the Expert shall control the resolution of any dispute referred in accordance with these procedures and Rules by exercising any or all of the powers expressed or implied herein. Specifically the Expert shall have the power to act inquisitorially in determining the dispute.'

This means the expert has wide powers including appointing advisers after having given to the parties ten days notice and an estimate of the costs. The rules also provide for the payment of interest and the authority to provide an interim decision where appropriate. The expert can be proactive in searching out evidence and is not restricted to those points raised by the parties themselves or their representatives. The guide notes to the rules are very useful; they provide information on the legal position of the expert in the performance of the expert's duties contained within the contract but do not remove the need for appropriate advice.

The annexes

The following annexes are provided within the expert procedures and set out standard terms, together with standard format letters and agreements that may be used in connection with the expert determination:

Annex A – expert's fees and expenses
Annex B – institution charges
Annex C – application for nomination of an expert by IChemE
Annex D – Form of agreement to appoint an expert
Annex E – Form of agreement to accept the appointment of an expert in accordance with the IChemE rules and procedures
Annex F– Form of expert's letter to the parties following appointment.

Chapter 13
New Engineering Form of Contract and the Adjudicator's Contract

Generally

This chapter looks not only at the second edition of the New Engineering Contract, renamed the Engineering and Construction Contract, but also briefly at its predecessor, including the consultative document. The changes to the dispute provisions referred to have special significance in that they have occurred over a very short period and are illustrative of the dynamics of our perception of adjudication and of the associated problems.

The consultative documents

The new engineering contract was first issued as a consultative document by the Institution of Civil Engineers in 1991. This contract was a significant departure from most existing standard forms of contract and was produced to provide a contract that would meet the changing demands of employers and overcome the perceived and actual deficiencies in existing standard form contracts. It was stated in the foreword to the consultative document that the proliferation of ad hoc, widely differing forms of contract was the cause of an increase in the number of disputes. Whether a new form on its own could overcome the deficiencies is doubtful but it does provide the focus for seeking other improvements. However, at a practical level take-up has been poor, if only because of the unknown.

It was a contract that would use core clauses, together with secondary option clauses, in conjunction with one of six main options which reflected the main contractual arrangements used. In addition, there was also a sub-contract for use in connection with the main form and this was similarly structured. The core clauses, as the name suggests, were applicable to all procurement options and were in nine sections, with section nine devoted to the procedures for dealing with disputes under the contract and for termination.

The guidance notes to the NEC consultative document[1] stated:

'Disputes in construction contracts are frequent and the ultimate means of resolution in most standard forms of contract is arbitration which in recent years has become both time consuming and expensive. In NEC, an intermediate stage of dispute resolution has been introduced in the form of adjudication. It is the intention that all disputes should be resolved by the adjudicator. However if either party is dissatisfied with the adjudicator's decision, he may refer the dispute to arbitration.'

NEC felt that the frequency of disputes and the associated time and cost of arbitration required another way forward and it is implicit in their approach that adjudication is better. The rationale for the NEC was that it could reduce disputes. The justification for these assertions was not made clear but helped shape Latham's thinking and the legislation that followed.

Section 9 contains provisions for both adjudication (clause 90) and arbitration (clause 91). The adjudicator under this contract carries out functions that are carried out by the engineer in many civil engineering standard forms, for example clause 66 of the ICE Conditions, and consequently has the advantage of separating the role of administrator from that of 'judge'.

First edition

Following the issue of the consultative documents, comments were received from a large number of organisations and the NEC form was used on a range of projects to provide feedback. The first edition of the New Engineering Contract was published in 1993 and although the principles and main features were unchanged, various drafting changes were made. Section 9 still embraced the basic concepts of adjudication and arbitration for dispute resolution. The underlying fundamentals remained the same but the provisions were completely redrafted to provide a different process. This substantial redrafting reflected the sensitive nature of the subject of dispute resolution and the growth in the awareness of the subject.

The presentation of the provisions is now in four clauses, which are headed:

Clause 90 – 'Disputes about an action taken by the Project Manager or the Supervisor'
Clause 91 – 'Disputes about an action not taken by the Project Manager or the Supervisor'
Clause 92 – 'The Adjudicator'
Clause 93 – 'Arbitration'.

Clause 90.1 changed in tone in that it drops the word 'disagree' from the provision but more importantly removes the right of the employer to dispute an action of the project manager (PM). The contractor only needed to believe an action of the PM was not in accordance with the contract and he could 'notify' as opposed to 'refer' it to the adjudicator and additionally the PM. Clearly, this was an attempt to defuse the situation at an early stage by removing an emotive word, reducing formality and maintaining communication with the PM. The clause also made it clear that the parties must implement the disputed action only so long as the contract has not been terminated. A more fundamental shift was that only the contractor could notify the adjudicator, whereas in the consultative document and in contracts in general both parties could submit disputes. The NEC took the view that as the PM was acting on the employer's behalf it was incongruous to allow the employer to challenge the PM's action or non-action.

The PM had two weeks from the notification to provide the adjudicator and contractor with information on which the dispute was based, and within two weeks of receiving this information the contractor could provide the adjudicator and PM with any other information. According to the guidance notes, information that became available after the commencement of the dispute was inadmissible and the dispute should not be widened to embrace matters not notified. The logic of this was clear in that to do otherwise could create an unfair situation. Not to take account of other matters can in practice often mean there is potential for a subsequent dispute to occur. As a comparison, clause 66 of the ICE Conditions of Contract when read with rule 4.1 of the ICE Arbitration Procedure 1997 can lead to the dispute referred being enlarged. Clause 90.2 provides a time frame that was missing at the consultative stage and also encourages good practice by requiring that information passes not only to the adjudicator but also to the other party. The adjudicator's decision is provided for under clause 90.3 and is essentially the same as provided in the consultative documents.

The first edition addressed a major shortcoming in that it provided for not only disputes on actions which had been taken but also for actions not taken and which it was believed should have been. Under clause 91.1, where the contractor believed the PM had not taken action expressly required under the contract, the PM could be notified accordingly and had four weeks in which to act. During this time if the PM chose not to act, it would be prudent to persuade the contractor of the validity of the non-action because at the end of this period the contractor had the right to notify the adjudicator within a further four weeks. Clause 91.2 stated that the contractor could in this notification give reasons why the PM should have acted and the PM had a further two weeks to respond.

If a dispute proceeded under clause 91, the adjudicator had to decide whether or not an action should have occurred and, if so, the PM was required to implement the action. In addition, an assessment of any additional cost and delay was to be made. The adjudicator had four weeks from receiving the information to give a decision, together with reasons.

The adjudicator was named in the contract data together with any replacement should this person resign or be unable to act. A replacement adjudicator was required to decide any outstanding disputes within the appropriate timescales, i.e. from the date of notifying the person to act as adjudicator. This required the employer to insert the name of the adjudicator before issuing the contractor with the invitation to tender. The choice of adjudicator was therefore at the employer's option. Although it was possible not to name the person in the contract data, provision was made for identifying the person who would choose the adjudicator if the parties failed to agree. If the employer identified an adjudicator in the contract data whom the contractor found unacceptable, this should be indicated when submitting the tender. The need to agree an alternative would not be an auspicious start and emphasises the necessity to select an appropriate person. Of course, the appointed or named adjudicator might not be able to perform the role for a variety of reasons and in these circumstances the parties had four weeks to choose another adjudicator, failing which the person named in the contract data would select an adjudicator whom the parties must accept.

There was no reference to the terms of the adjudicator's appointment other than that payments were to be shared equally by the parties. However, the guidance notes expected that the adjudicator usually would be appointed under the NEC professional services contract (it should have read adjudicator's contract) (see the commentary on the adjudicator's contract later in this Chapter).

The joinder provision was contained in clause 92.3 and remains as it was at consultative stage, except for the use of 'notify' in lieu of 'refer'.

Arbitration was provided for under clause 93 and was simplified. Arbitration was available to either party if the adjudicator failed to give a decision within four weeks, as provided in clause 92.1 or if a party disagreed with the adjudicator's decision. But arbitration could not be commenced unless the dispute had first been referred to the adjudicator. This was problematic, in that only disputes involving the PM were notifiable and therefore some disputes would appear to have had no competent forum for settlement within the contract provisions and would proceed to litigation. The timescales were changed slightly and such a reference was required within four weeks from when the adjudicator gave a decision or should have given a decision. Like many aspects of a contract, the four week period, could be extended by agreement if the adjudicator requested a longer period for providing a decision in a complex dispute. The power to challenge any decision of the adjudicator was available. The giving of reasons for the decision was laudable in the context of fairness and openness but could provide a stimulus for challenging the decision that might not have otherwise arisen.

The provisions dealing with the appointment of the arbitrator and the procedure to be used in arbitration were curtailed and clause 93.2 simply mentioned, 'A dispute referred to an arbitrator is conducted using the arbitration procedure'. This procedure was to be identified in section 9 of the contract data and should be inserted by the employer prior to entering into a contract. The guidance notes identified what were considered suitable procedures for the UK and overseas.

Subcontract

The first edition of subcontract contained the same basic changes from the consultation document and reflected the procedure in the main contract for dealing with both adjudication and arbitration. The only difference concerned the matter of joinder and clause 92.3 enabled a subcontractor to refer a dispute arising under a sub-subcontract which was also a dispute under the subcontract and for the adjudicator to decide the two disputes. Clause 92.4 also provided that where a dispute under the main contract involved the subcontract, the contractor could require the subcontractor to provide information as requested.

Engineering and Construction Contract

The Engineering and Construction Contract is the second edition of the NEC contract but was renamed following a recommendation by Sir Michael Latham in his report *Constructing the Team*. The second edition was published in November 1995 and contains some significant changes as regards the resolution of disputes, indicating yet again that our understanding of dispute resolution is still evolving in important ways and that our knowledge is still far from complete. This contract does not comply with the requirements of the HGCRA and a subsequent addendum has been published and is discussed separately later in this chapter.

Section 9 remains the relevant part dealing with disputes but contains many substantive changes and differences in format. A major change was to replace the arbitration clause with a provision dealing with a reference to a tribunal. This tribunal may be either arbitration or the courts, although it is not restricted to these.

The dispute resolution procedures contained within the second edition, like the earlier edition, have been the subject of much debate, and concern remains that they could prove unworkable.

Submission of dispute

Clause 90.1 provides that any dispute shall be submitted to the adjudicator. The clause contains an adjudication table which sets out which disputes may be submitted, who may make the submission and the various time constraints involved.

Disputes concerning the actions or non-actions of the PM are submitted only by the contractor and the second edition adopts the same position as the first edition, except that the argument about whether this was an obligation or a right has been settled in favour of the former. Furthermore, an addition has been made to cover 'other' disputes and these may be submitted by either party. Thus, a significant deficiency in the first edition is plugged and adjudication is now a mandatory first tier of the dispute resolution process. Under this form all disputes are settled by the adjudicator, potentially extending to legal issues and matters between the parties that might emerge later, for instance latent defects arising outside the defects liability period.

The time constraints have been changed to provide an opportunity to resolve disputes without the need for reference to the adjudicator, yet at the same time strengthening the provisions. Where the dispute concerns the PM, the contractor must notify the PM within four weeks of becoming aware of the action or non-action. The dispute may then be submitted to the adjudicator between two and four weeks after this notification. The date of action is self-evident but the date when an action should have been taken is not always apparent and therefore whether notification is within time could become a matter of contention. If there is any doubt, pragmatism should prevail and err towards accepting the notice as valid, unless it is to a party's advantage to create confusion.

The initial two-week period during which the dispute cannot be submitted to the adjudicator provides an opportunity for the PM to resolve the dispute directly with the contractor. This is a sensible approach and it is likely that many matters will be resolved at this stage.

Where a dispute does not involve the PM, the aggrieved party must notify the other party and the PM and it may then be submitted to the adjudicator between two and four weeks after this notification. The two-week lapse will similarly provide the parties with an opportunity to settle the dispute before it is submitted to the adjudicator.

The contract does not provide for what happens if either party fails to instigate the procedure within the timescale and still wishes to dispute the matter. Does such a failure mean that a remedy is denied? The tribunal can only review decisions of an adjudicator or where the adjudicator fails to make a decision within the time provided. The question is whether the parties would then be required first to use an adjudicator in an attempt to resolve the dispute regardless of their failure to comply with the appropriate timescales. Therefore, although there is a timetable for referring disputes to the adjudicator, which should be adhered to, it is probably worthwhile still using this provision even if out of time. If this course is not followed there remain the interesting questions as to whether a party could become out of time for the commencement of arbitration and could then proceed to litigation, notwithstanding clause 93 provisions making adjudication a conditional precedent. However, it is clear that if the procedure is not followed the parties must proceed as if there were no dispute.

The adjudicator

The adjudicator referred to in clause 92.1 is the person named in the contract data and follows the procedure adopted by the first edition. It is stated that the adjudicator is independent and does not act as an arbitrator. The adjudicator may, therefore, be acting in a quasi-judicial capacity and will not be given the immunity from an action in negligence, that is extended to arbitrators at common law and under statute, unless the adjudicator's contract restricts liability (see adjudicator's contract later in this chapter). Independence is also an important issue and the employer should not name an adjudicator where a conflict of interest exists. It is of course possible that a conflict of interest may arise after appointment and in these circumstances the adjudicator should remove the conflict or at the very least disclose the conflict. One such action may be to resign as the adjudicator on the contract. This is generally better than continuing, even where the parties following disclosure are not unhappy.

Important additional powers have been given to the adjudicator, which include the power to review and revise any action or inaction of the PM or supervisor related to the dispute. This issue was not dealt with at the consultative stage and was only marginally improved on in the first edition.

Clause 92.2 applies where the adjudicator resigns or is no longer able to act. The parties then select another adjudicator and failure to do so within four weeks will mean that either party can then ask the person named in the contract data to choose a new adjudicator. The person chosen must then be accepted by the parties. The date of appointment of the new adjudicator is taken to be the date of the submission of any outstanding disputes and the prescribed timescales will run from this date in respect of these disputes.

The replacement adjudicator is appointed as an adjudicator under the NEC adjudicator's contract (clause 92.2). By contrast, there is no reference to the terms of the original adjudicator's appointment. However, the guidance notes still anticipate that usually this adjudicator will be appointed under the NEC adjudicator's contract.

The adjudication

Clause 91.1 states that when notification of a dispute is given it is to include information which the adjudicator is to consider. Although the information provided at this stage should be as full as possible, further information or indeed amendments to the submission may be made within four weeks from the original submission. The other party may within this timescale submit a response to the submission but difficulties may be encountered because of the ability of the instigator to amend a submission. Similarly, the adjudicator may also request further information and may seek to extend this period by agreement. The purpose of adjudication is to achieve quick settlements of disputes, so any extension should be avoided wherever practicable.

The adjudication provisions contain the usual joinder clause (91.2), which provides for the subcontractor to become a party to the adjudication where the dispute under the main contract is concerned with subcontract work. However, unless the subcontract similarly provides, the subcontractor could resist such action. The published subcontract makes such provision and the subcontractor would be so obliged where this subcontract was used.

The adjudicator's decision

The adjudicator's decision, together with reasons, is given to the parties and the PM within four weeks from the end of the period for providing information or such extended time as is agreed. The decision is final and binding unless it is subsequently revised by the tribunal. Until the decision is made the project should proceed and its progress should not be impeded in any way because of the dispute.

In arriving at a decision, which involves any assessment of cost or delay, the adjudicator must do so in the same way as for a compensation event under clause 60. If an amount due is corrected in a later certificate following a decision of the adjudicator, interest at the rate stated in the contract data is also payable in respect of the difference.

The form that the decision and reasons take is very much up to the adjudicator but must be scrupulously drafted. In this regard, the guidance notes contain some very helpful advice.

Clause 92.1 expressly states that the adjudicator's decision is '. . . enforceable as a matter of contractual obligation between the parties and not as an arbitral award'. This provision appreciably strengthens the adjudicator's decision by making its status clear and enabling the parties to treat non-compliance with it as a breach of contract. However, this is rather undermined by the fact that the adjudicator's decision can subsequently be challenged in the specified tribunal. Also, there is the issue of whether a party sued for non-compliance with an adjudicator's decision would be able to raise a defence related to the inappropriate conduct of the adjudicator during the adjudication.

Review by the tribunal

A major change in the second edition is the introduction of review by the tribunal rather than simply reference to arbitration. This is a major change in that the employer now has to identify the tribunal in the contract data. This could be arbitration, the courts, expert determination, dispute review board or any other such forum, but should never be a named individual. In practice, arbitration and the courts are the most likely choices. Where arbitration is stated to apply, an optional statement is also required in the contract data, namely which arbitration procedure will apply. The guidance notes identify various procedures that are applicable to civil engineering, building and overseas work, and the employer should identify the most appropriate procedure for the work involved.

Clause 93 now provides for either party to refer any matter, previously referred to the adjudicator, to the tribunal for review. This applies to any decision of the adjudicator and to any dispute which the adjudicator fails to decide within the stipulated time. In order to preserve the right to this review, the dissatisfied party must give notice of intention to refer the matter disputed within four weeks. The four-week period runs from the date of notification of the adjudicator's decision or the date on which it should have been given if this is earlier. The reference in clause 93.1 to 'the time provided by this contract for this notification...' must also be intended to cover agreed extensions to the four-week period but this is not especially clear in the words used and could lead to further dispute. Although the above timescale must be complied with, the review by the tribunal cannot take place until after completion or the earlier termination of the works. This means that the binding nature of the adjudicator's decision until it is overturned by the tribunal has great significance. Failure to comply with the adjudication timescale was discussed previously but in order to be certain that the option of a tribunal remains, a notice of referral to the tribunal must be given within the specified time limit.

The tribunal is given wide powers and may review or revise any decision of the adjudicator, or action or inaction on the part of the PM or supervisor which is related to the dispute before it. Clause 93.2 makes it clear that during the course of the review, the parties may introduce any other information and they may advance any other argument, whether or not presented to the adjudicator.

Early warning

A novel provision of the NEC contract is the requirement under clause 16 to give early warning of certain events that could increase the price, cause delay or impair performance of the works. This is a laudable attempt at dispute avoidance by trying to bring the parties together to resolve their own problem. But, as referred to in the guidance note, the sanction in the event of failure to give early warning, is to reduce payment; this may actually increase the potential for disputes.

Subcontract 2nd edition

The second edition of subcontract broadly reflects the procedures in the main contract that deal with the settlement of disputes. Again, the main difference concerns the matter of joinder. Clause 91.2 enables a subcontractor to refer a dispute arising under a sub-subcontract which is also a dispute under the subcontract and for the adjudicator to decide the two disputes. Clause 91.3 also provides that where a dispute under the main contract involves the subcontract then the contractor can require the subcontractor to provide information as requested. Additionally, clause 93.2

makes provision for the challenge of a subcontract dispute that has been referred to the main contract adjudicator. The clause provides for such a decision to be referred to the tribunal, as long as the four-week timescale is complied with but in any event, not before completion of earlier termination.

NEC Engineering and Construction Contract and Subcontract Addendum 1998

The NEC panel approved amendments to the NEC contracts so that they would comply with the HGCRA 1996. These amendments together with guidance notes were published by the ICE as an addendum in April 1998. The addendum sets out Option Y(UK)2: The Housing Grants, Construction and Regeneration Act 1996 and is for use where this Act applies and where the parties do not wish the provisions of the Scheme for Construction Contracts to become implied terms of their contract. The new provisions are, as they say, 'optional' and need not apply for other reasons, such as the contract is outside the UK or it is for works not covered by the Act. Where the option is to be incorporated into the contract, the contract data should include a statement to this effect, as contained in the notes for guidance.

Option Y(UK)2 contains amendments to cover both the payment and the adjudication requirements of the Act. Where work falls under the HGCRA and the parties do not wish to rely on the Scheme for Construction Contracts they can incorporate the addendum. The clauses relevant to adjudication are Y2.5 and 2.6, and clauses 90 and 91.1 of the contract are deleted and replaced with new provisions in order to retain the principles of adjudication and to comply with the HGCRA.

Addendum clause Y2.5

The optional new clause 90 is headed 'Avoidance and settlement of disputes' and contains the procedure to be followed in the event of a problem. The procedure sets out clearly when a dispute arises and follows broadly the ICE approach of first resolving dissatisfactions before creating a dispute by definition.

Dissatisfaction and a dispute

Clause 90.2 provides that where one party is dissatisfied about some matter under the contract, they notify the other party and then meet to resolve the issue within two weeks of the notification. The purpose of this procedure has been discussed in Chapter 12 but in essence it is a process that delays the right to adjudicate in the hope that most matters will be resolved more readily. This position is reinforced by clause 90.4 where it is made clear that a dispute does not arise, and hence adjudication cannot begin, until a notice of dissatisfaction has been issued and this has not been resolved within four weeks of this notice. This is a sensible procedure but it does give rise to a possible complication in that any matter of dissatisfaction must be raised within four weeks of a party becoming aware of the event from which the dissatisfaction emanates.. The question may then arise of what happens where a notice of dissatisfaction is not given within this four-week period but one party clearly wishes to pursue the issue. Common sense says that a dispute has arisen and the adjudication process should be activated but a dispute by definition does not exist and one must exist before adjudication can start.

This raises the following questions. Does this mean that a right to adjudication is denied by the contract because of the way it defines a dispute? If so, is this intentional? Can the party proceed directly to the tribunal? It would appear not, because an adjudicator must have acted or failed to do so, before this can happen. Does the failure to give notice, therefore, deprive the party of all remedies?

It is difficult to give clear answers to these questions but the failure to comply with the four-week period may not necessarily be considered a condition precedent to adjudication. The attempt to define disputes may be avoided or redefined and hence a party may have maintained a right to adjudication according to the contract. Adjudication may still be commenced under the contract, notwithstanding that this is not the intention. It may be in the best interests of the parties to follow this route. The alternative view is that the definition of a dispute is too restrictive and that a dispute does exist for which adjudication is not provided for in the contract. Under these circumstances the contract provisions would be supplemented by implying the adjudication terms contained in the Scheme for Construction Contracts.

Notice of adjudication and the adjudicator

Either party has the right to give notice of his intention to start adjudication but there must be a 'dispute' to refer. After giving this notice of intention, the referring party has seven days to refer the dispute to the adjudicator together with the information to be considered. It is intended that the adjudicator is named in the contract and therefore the party will know whom to notify. Any further information to be considered by the adjudicator, which will include any response by the other party, must be submitted within 14 days of the referral.

Clause 90.8 requires the adjudicator to act impartially and to take the initiative in ascertaining the facts and the law and reflects the requirements of items (e) and (f) of section 108(2) of the HGCRA. Clause 90.12 provides the adjudicator with immunity, except when acting in bad faith, and this provision is inserted to comply with section 108(4).

The adjudicator is still referred to in clause 92 and this clause remains unaltered except for some small drafting changes referred to later in this chapter.

Decision

Clause 90.9 sets out the times for reaching a decision and again they reflect the 28 days from referral or as otherwise agreed after the dispute has arisen subject to the adjudicator's ability to extend time by up to 14 days with the consent only of the referring party. This mirrors the requirements of items (c) and (d) of section 108(2) of the HGCRA. Until a decision has been formulated, the parties must continue with the works as if no dispute existed.

Decisions of the adjudicator must be given with reasons and this is a significant departure from many other forms of contract that do not require reasons. Any decision of the adjudicator is binding until the dispute is finally determined by the tribunal or by agreement of the parties.

Combining procedures

Clause 91 is relabelled and 91.1 is replaced to enable subcontractors to attend meetings to resolve matters of dissatisfaction which are common with those under the main contract. Other small drafting changes are made to clauses 91 and 92 and the various references to the adjudicator 'will settle(s)' or 'settled' have been changed and relate now to the adjudicator's 'decision'.

Subcontracts have a similar addendum for use in connection with the form of contract. The wording is the same as the main form with the necessary changes having been made. The major change is to include a new clause 91.3, which requires attendance of the subcontractor at meetings, where the main contract so provides, to resolve matters of dissatisfaction. The renumbered clause 91.3 (now 91.4) permits the subcontractor to have joint adjudication if allowed in the main contract.

The future

The second edition certainly makes significant improvements to the earlier versions but the absence of a fully binding decision by the adjudicator may encourage the decision to be challenged, especially as other evidence can be provided at the tribunal. The precise role of the adjudicator, where Option Y(UK)2 is not operative, will require clarification and in the meantime will inevitably lead to problems. How effective the latest provisions will be is a matter of conjecture and only time will provide a clear view. The matter of using 'dissatisfaction' to avoid a dispute may also require to be revisited, a problem shared with the ICE Conditions of Contract as amended.

NEC Adjudicator's Contract

As the NEC forms of contract require the appointment of an adjudicator, it was considered necessary to have a standard form of adjudicator's contract. The NEC issued a consultative version in 1992 and, following feedback, produced the first edition of the Adjudicator's Contract in 1994. This was followed by a second edition of the Adjudicator's Contract together with a separate booklet containing guidance notes and flow charts in June 1998.

The Adjudicator's Contract was designed to be used with the NEC family of contracts including the Professional Services Contract. The guidance notes for this contract also indicate that it can be used with other forms of contract but care must be taken to avoid any conflict between this contract and the main contract. Clause 1.7 of the Adjudicator's Contract establishes that priority is given to this contract in the event of such a conflict occurring, except of course, where the Scheme would apply. Nevertheless, it is far preferable to avoid any conflict in the first place and this should be done by making suitable amendments to whichever contract requires to be changed.

The Adjudicator's Contract has been kept fairly simple, comprising two separate forms of agreement, the conditions of contract and the contract data, a total of six pages.

Form of agreement

The second edition of the Adjudicator's Contract provides two separate forms of agreement, one where the Scheme of Construction Contracts does not apply, and the other where the Scheme does apply.

The first form of agreement in the contract is where the Scheme does not apply. This agreement simply requires that both the parties to the main contract, together with the adjudicator, are named and their addresses stated. It contains two very brief provisions that set out that the parties are appointing the adjudicator to act under the conditions of contract and contract data, and that the adjudicator accepts such appointment and will carry out the duties described in the adjudicator's conditions of contract. This form is signed by the parties and the adjudicator.

The second form of agreement in the contract is where the Scheme does apply and contains a headnote to this effect. The requirements regarding names, addresses and signatures are the same but this form of agreement contains three brief provisions. The first two simply contain a reference to the Scheme for Construction Contracts and that the appointment, acceptance and duties of an adjudicator are also in accord with these provisions. The third provision is additional and indicates that the words 'and the Scheme of Construction Contracts' should be added to references to 'the contract between the parties'. Furthermore, it provides for the adjudicator to apportion fees and expenses between the parties, rather than in equal shares.

When an agreement has been entered into where it is not intended that the

Scheme will apply, but later it becomes known that the Scheme will apply (because the contract does not comply with the HGCRA), then the other form of agreement should also be entered into. Adjudicators should take care to ensure that they do not accept an appointment without making known the terms of that appointment, whether by signing a form of agreement or otherwise.

Conditions of contract

The conditions of contract comprise five brief sections which cover:

- general matters including definitions
- adjudication process and decision
- payment of the adjudicator's fee
- risks – indemnity
- termination of adjudicator's appointment.

The first edition contained an additional section headed 'title' and part of this is now contained elsewhere, with the remainder considered unnecessary.

Section 1 – general

Clause 1 covers a range of general matters that are self-explanatory and generally require little or no comment. Points that are worthy of comment are as follows.

The adjudicator shall act impartially and where the adjudicator becomes aware of anything that may give rise to a conflict of interest or render him unable to act, the parties should be informed. Subclause 1.1 together with the form of agreement means that the adjudicator is required to carry out any procedural duties set out in the adjudicator's contract, and by reference, the contract between the parties and where applicable the Scheme for Construction Contracts.

Clause 1.4 – This sets out what are the recoverable 'expenses' and covers any expenses incurred by the adjudicator in carrying out the adjudication and includes payments made to others for help on the adjudication. Under clause 2.2, the adjudicator may obtain help from others to reach a decision but the decision must be that of the adjudicator. No prior approval from the parties, or indeed notification, is required before the adjudicator seeks help from any other person. This is rather surprising because it leaves the adjudicator with total discretion to act, without the parties knowing that it is occurring and without knowledge of the possible costs involved. It would seem that a prudent adjudicator would keep the parties informed, notwithstanding that there is no obligation to do so under the contract.

Clause 1.9 – This requires that all communications required by the contract shall be 'in a form which can be read, copied and recorded'. According to the guidance notes the clause is intended to cover 'a letter sent by post, a telex, cable, electronic mail, facsimile transmission, and on disc, magnetic tape or other electronic means'. This wording, in the age of information technology, is preferable to 'in writing' and should avoid many misunderstandings as to what constitutes the correct form of communication.

Any communication has effect once received at the last notified address. This means that the address stated in the form of agreement is applicable unless a subsequent change is notified. This applies equally to the employer, contractor and adjudicator and care should be taken to inform the others if an address should change.

Section 2 – adjudication

The process of adjudication is covered in clauses 2.1–2.4. These provisions have undergone change from the earlier edition and detailed procedures are not included as they are in the NEC forms of contract. The adjudicator is, therefore, required to act in accordance with these procedures or, where applicable, those of the Scheme and any failure to do so means that his decision may be successfully challenged. In addition, the adjudicator may obtain help from others and this has been discussed earlier.

The adjudicator's decision and any information relevant to the adjudication are confidential and can only be used by those with a proper interest. Although this may extend beyond the parties, it is by no means a public decision and no one else can use the decision to further another case.

Section 3 – payment

Clauses 3.1–3.7 set out the payment details but make no provision for a retainer to be paid to the adjudicator. This means that an adjudicator is only paid if a dispute is referred. Unless the Scheme applies, the parties are equally responsible for the adjudicator's fees and expenses, jointly and severally liable. Therefore if one party defaults by not paying, the adjudicator can recover the outstanding amount, together with interest, from the other party. As each decision is notified to the parties the adjudicator shall invoice the parties for one half of the fees and expenses, including any value added tax, which shall be paid within three weeks of the date of invoice or such other date as is inserted in the contract data. The three-week payment period runs from the invoice date and not the date of receipt. Consequently, the actual payment period will often appear much shorter. Interest is payable on a late payment. Nowadays interest may be recoverable, if not mentioned, under the Late Payment of Commercial Debts (Interest) Act 1998.

In circumstances where one party has had to pay the defaulting party's share of the adjudicator's fees and expenses, these, together with interest compounded annually, are said in clause 3.5 to be recoverable from the defaulting party. As the adjudicator's contract is a tripartite contract, this clause would provide an enforceable right.

Where the Scheme applies, the adjudicator apportions the fees and expenses and this is taken under clause 3.1 to be otherwise agreed. Once an adjudicator has apportioned the fees and a party defaults in payment, there is the strange position of seeking redress against the other party, possibly the party who had no fees apportioned to them.

Section 4 – risks

The adjudicator and his employees are given protection under clause 4.1 for anything they do or fail to do in the process of the adjudication unless as a result of bad faith. This means that there is no redress for a party against the adjudicator unless the party can show that they have suffered loss because of bad faith. A definition of 'bad faith' is difficult but may be taken to mean acting with deliberate intent to provide an inappropriate outcome. Those who assist the adjudicator but are neither employees nor agents are not given any measure of protection from legal action.

Indemnity clauses in contracts are not uncommon; subclause 4.2 reads as follows:

> 'The Parties indemnify the Adjudicator, his employees and agents against claims, compensation and costs arising out of the adjudicator's decision unless ... made in bad faith.'

The guidance notes emphasise that it is important that the adjudicator be given protection from actions arising out of the performance of his role. The NEC takes the view that for a successful adjudication to take place, the adjudicator should be free of the concern that a disappointed party would seek redress from them.

The indemnity referred to above protects the adjudicator from the claims of third parties. This does not mean a third party who can establish a duty of care and reliance in tort cannot successfully sue the adjudicator in negligence, but where this occurs the parties would be required to reimburse any loss. It probably goes further because there is no specific reference to legal proceedings and therefore the adjudicator or insurer could settle an amount and claim this from the parties. This looks potentially nasty for the parties but should be a rare event.

This protection of claims from third parties recognises that the adjudicator is not acting in a quasi-judicial capacity but it would seem that as far as the parties are concerned, that is what is almost achieved.

Section 5 – termination

On the surface, this is a straightforward and sensible provision which enables the parties to agree to terminate the appointment of the adjudicator.

If the parties seek to terminate the appointment they are required to notify the adjudicator and under clause 1.9 this communication has to be in a form that can be read, copied and recorded, and sent to the last notified address. Such an action is unlikely to be common and were it to occur, how would the parties come to the point of agreeing that an adjudicator should be removed? It is clear that one party on their own have no such right and would need to persuade the other.

The adjudicator's own appointment may be terminated where he is prevented from carrying out the adjudication because of conflict of interest or if unable to act. This might cover events such as:

- being physically prevented by not being provided with access
- obstruction caused by both of the parties
- non-compliance by both parties
- a dispute arises outside his area of competence.

Termination may also occur because of failure of a party to pay an amount within five weeks of the date when payment was due. As the payment can be made up to three weeks from the invoice date, a period of eight weeks must elapse before an adjudicator can terminate for non-payment. This could be substantially longer if a longer period for payment is inserted in the contract data.

Unless the adjudicator's appointment is terminated in one of the ways described above, it will terminate by the expiration of time. Clause 5.3 states that 'the appointment terminates on the date stated in the Contract Data'.

An adjudicator's appointment could be terminated before all disputes are settled. The point at which no further disputes can be referred to the adjudicator is uncertain and therefore a realistic time needs to be inserted. This does not apply to the Scheme because the adjudicator is appointed for a specific dispute.

Contract data

This is a short section of the contract that sets out:

- what is the contract between the parties – so that the adjudicator knows under what contract any adjudication will occur
- the law governing the adjudication contract – this contract may be used for international works
- the language of the contract – because communication must be in this language
- the adjudicator's hourly fee is to be stated – this hourly fee applies to all the time spent on the adjudication, including all preparation and time spent travelling in connection therewith. This does not include the cost of the adjudicator consulting others and that is an additional cost to the

parties. Payments to such persons are part of the adjudicator's claimable expenses (see clause 1.4) and are recoverable from the parties
- an interest rate is inserted – this rate will be applicable if the adjudicator's fee is not paid within three weeks of the date of invoice or such other date as is stated in the optional statement in the contract data. This may now reflect the Late Payment of Commercial Debts (Interest) Act 1998
- the currency in which the adjudicator's fee is payable can be inserted – this covers the situation whereby the adjudicator is from a country different from that of the construction contract. If no insertion is made, the same currency as is applicable to the construction contract will be used
- the termination date of the adjudicator's appointment – this is referred to in clause 5.3.

Professional Services Contract

The Professional Services Contract (PSC) has been developed by the ICE as a part of the NEC suite of contract documents. It is a standard form of contract for use with consultants providing professional services in connection with engineering and construction works. Although designed as an NEC document, its use is not limited to NEC contracts and it may be used with other main contract forms.

The PSC utilises the core clauses for disputes and these are contained in section 9. In a similar way that the main contract provides for subcontract disputes that form part of the main dispute to be dealt with by the same adjudicator, clause 91.3 provides for disputes arising with subconsultants.

Chapter 14
Model Rules for Adjudication

Generally

As adjudication becomes an integral part of the dispute resolution process within the construction industry, the need for protocols will continue to emerge. This chapter considers adjudication rules produced by bodies independent of contract authoring and publication. The Centre for Dispute Resolution (CEDR), among others, recognised this need and considered a draft set of Model Rules for Adjudication in May 1996. At this time there were still a number of unresolved matters: whether adjudication should be an interim or a fully binding procedure, the nature and extent of training for adjudicators and the fee structure to be adopted for CEDR services.

These and other issues were the subjects of continuing debate and it was not until late 1997 that the 'final' draft version was produced. Even then the rules were not published by the CEDR until May 1998 and contained several changes, especially concerning time periods. Although CEDR is concerned with disputes from other industries, the CEDR Rules for Adjudication were designed for construction adjudications and those of a domestic nature. If the dispute related to an international contract and the CEDR adjudication procedure was to be used, it would need to be amended to cover language and applicable law and jurisdiction. The CEDR has produced clauses for use in this context and these are contained within the guidance notes.

CIC Report on Appointment of Adjudicators

Following publication of the HGCRA 1996, the Construction Industry Council (CIC) published in October 1996 the *Report on Appointment of Adjudicators*. This report was compiled from the views of CIC members and other relevant bodies concerned with arbitration and the law. It reviewed the arrangements, current at that time, for the appointment of adjudicators and the present contractual adjudication provisions. It outlined different options for the appointment of an adjudicator and how the CIC was positioning itself so that it might be an appointing body under the Scheme for Construction Contracts. In the event the latter became a non-issue. The idea of approved appointing bodies was not maintained in the final version.

Despite there being no reference to adjudicator appointing bodies in the Scheme for Construction Contracts, the CIC continued with its work relating to adjudication and produced the first edition of their Model Adjudication Procedure in February 1998.

The CEDR, CIC and the Official Referees Solicitors Association (ORSA) adjudication procedures are discussed below to assist in selecting an appropriate adjudication procedure. It would seem likely that most parties will find themselves drawn automatically to the procedure that is linked to their contract but others may choose an alternative procedure. Sadly, this will create a degree of confusion and it is to be hoped that even if a number of adjudication procedures continue to exist, the CIC Model Rules will become the base from which they are drafted. Already there is much similarity in many of the rules.

CEDR Rules for Adjudication

The draft model rules were in two parts, part one dealing with the appointment of an adjudicator and part two providing rules of conduct. In addition the rules cover a form of application for the appointment of an adjudicator, together with an adjudicator's form of agreement and alternative model clauses for inclusion in a contract for the appointment of an adjudicator if there is a dispute. This draft procedure formed the basis of the procedure published in 1998 but a number of differences emerged as the debate on adjudication developed, especially in connection with the HGCRA.

These rules are not dissimilar to the approach used by the NEC (discussed in Chapter 13) and the CIC (discussed later) and there is every likelihood that a fairly common approach to adjudication will eventually emerge from the main players. However, at present we are continuing to witness a 'go-it-alone' approach from the various interested bodies.

Model clauses

CEDR provides two optional model clauses that provide for 'Any dispute arising under this Agreement...' to be referred to adjudication conducted in accordance with their Rules of Adjudication, whether or not an adjudicator has been named in the contract. Alternative B is the briefest of provisions, whereas Alternative A refers to the adjudicator's decision being in writing and promptly implemented. Additionally, it provides for the right to issue a notice within 60 days if a party is dissatisfied with the decision. The parties may agree any other period if they wish; the 1996 draft stated a period of only 30 days.

This ability to give notice that they are dissatisfied with an adjudicator's decision enables a party to pursue the dispute in arbitration or litigation as provided for in accordance with the contract. Despite such a challenge, the decision of an adjudicator is binding and must be implemented until the dispute is finally resolved in arbitration or litigation. If a notice is not served within any prescribed period the adjudicator's decision becomes final and binding and the parties give up any right of counterclaim, set-off or abatement. This giving up of rights can only refer to the disputed matter, although the clause does not state this expressly.

Alternative A also provides an optional provision which, if adopted, makes the adjudicator's decision final and binding until practical completion, notwithstanding the issue of a notice of dissatisfaction with the adjudicator's decision. This in effect provides a stay of other proceedings and although this may be desirable it is by no means necessary.

If a dispute occurs under either of these model clauses, the Rules of Adjudication become operative.

Appointment of an adjudicator and the reference to adjudication

Part 1 sets out the rules governing the appointment of an adjudicator and the reference to adjudication. These rules enable the parties to agree the appointment, failing which either party can apply to CEDR to make an appointment, by setting out the relevant details and providing a brief statement of redress that is sought. This procedure is facilitated by the use of a standard application form and requires a fee to be paid. CEDR is then obliged to appoint an adjudicator within seven days and there is no express right of objection to the appointee. If either party believed there to be a good reason why the adjudicator should not act, they could still make representations to CEDR or show the adjudicator why the appointment should not be accepted or have the adjudicator prevented from acting by court order for cause. Provision is also made whereby the parties are required to enter into a contract with each other and the adjudicator under the CEDR standard form of agreement but are bound by the adjudication rules whether they have signed the agreement or not.

The CEDR Adjudication Agreement is a straightforward two-page document which covers a number of points that are also dealt with elsewhere (see Chapter 13), except for the issue of confidentiality. Clause 4 states:

'The adjudication and all matters connected with it are and will be kept confidential by the parties except insofar as is necessary to enable a party to implement or enforce the decision of the Adjudicator or for the purpose of any proceedings subsequent to the adjudication.'

As both parties and the adjudicator sign this agreement they become bound by this clause. If confidentiality is important, and it usually is, it is necessary to ensure this agreement is entered into because the rules themselves do not cover this point.

Section C of part 1 also enables CEDR rules to be used where the parties have named the adjudicator in the contract or have subsequently agreed one without the services of CEDR. In this instance the party referring the dispute must give notice to the adjudicator together with all relevant details including names, nature and details of the dispute and the redress sought. The adjudicator should be required to enter into an agreement and in so doing should be confirming that he will act and in accordance with CEDR rules.

The date of referral of a dispute is an important issue and section D of part 1 sets out that it is either the date when CEDR appoints the adjudicator or when the adjudicator's confirmation to act is given to the parties, whichever is applicable. The first of these is clear but the confirmation date is open to debate and may require a notice quite separate from the agreement.

If the appointed adjudicator cannot act for any reason the parties should try to agree a replacement within seven days, failing which CEDR will, on application from either party, make a replacement appointment within seven days of receiving notice (rule 14 part 2).

Rules of conduct

Adjudicator procedure

Once an adjudicator has been appointed, part 2 governs the conduct of the adjudication. Under the rules of conduct the adjudicator may take the initiative in ascertaining the facts and the law and has wide discretion as to the procedure to be adopted for resolving the dispute. This means that the adjudicator is not bound by any rules except those provided for in the CEDR Rules of Adjudication, including rules of evidence. The adjudicator can also hold whatever meetings he believes necessary but cannot take account of statements made without making them available to the parties for consideration.

The adjudicator has wide discretion to vary any of the time limits under the rules of conduct except the time limit for the adjudicator's decision to be made, set out in clause 9. The reason this discretion does not apply to clause 9 is that this clause contains the essential time requirements of the HGCRA.

The adjudicator has great freedom in the conduct of the adjudication so long as he/she is impartial and acts in good faith. Both CEDR and the adjudicator are free from liability for their actions under the provisions unless the act or omission was in bad faith.

Unless the adjudicator establishes within two days that no written submissions are required or that a preliminary meeting will be held within seven days of the referral, the initiating party must submit to the adjudicator and the other party, a statement of the issues in dispute and relevant facts, together with any relevant documents. This must be done within four days of the referral and the other party may then respond within a further seven days of receipt. In the first draft, each statement, as opposed to other key documents, was restricted to 20 pages of text on A4 paper and although this has been dropped it provides a useful guide as to acceptable practice. The only reference now is in the guidance notes which says that the documentation should be kept as brief as possible. Although this is not mandatory, participants should try to achieve this otherwise one of the major benefits of adjudication will be lost.

The adjudicator may request other information, visit the site and conduct proceedings in any way to establish the facts and the law. The parties must act promptly when the adjudicator makes a request of them and a failure on their part to comply with the requirements of the adjudicator does not prevent the adjudicator from giving a decision.

The adjudicator may feel it necessary to obtain specialist advice but before doing so the consent of the parties must be obtained. This is sensible and provides the possibility of a veto on the use of such help. Although other adjudication procedures require notice to be given they do not require consent – see for instance the ICE Adjudication Procedure.

Adjudicator's decision

The decision, which is not an arbitral award, is required as soon as possible, that is 14 days from the referral but not later than 28 days from referral. The rules permit this period to be extended by up to 14 days with the consent of the referring party or as agreed by the parties. In practice, there is nothing to prevent consent being given to a longer period but this may defeat the purpose of the adjudication. However, in effect, this is what happens if mediation is sought before the adjudicator's decision because the period is automatically extended. Careful consideration should be given to a request from an adjudicator for an extension of time for arriving at a decision. If time is not granted this may prejudice the form of decision and may not be in the best interests of the parties. The adjudicator is still obliged to reach a decision that is appropriate to the circumstances. The adjudicator's decision may be given with or without reasons and this is a matter of discretion.

In reaching a decision an adjudicator has the power to review and revise any decisions relating to the dispute made under the contract, with the exception of those specifically excluded.

The parties are jointly responsible for the fees of the adjudicator, together with expenses and any fees and expenses of specialist advisers, but the adjudicator has discretion to apportion the fees as part of the decision. The adjudicator may make the publication of the decision subject to full payment of all fees and expenses. Whether this complies with the HGCRA is discussed in Chapter 13.

CEDR mediation

Under Rule 15 it is possible for the parties to agree to implement the mediation provision and this may be done at any stage prior to the issue of an adjudicator's decision. Agreement is necessary because the process is non-binding and works on co-operation. Only when a settlement is achieved and confirmed in writing in a form acceptable in law can it be binding. If a solution is to be made final and binding, advice should be taken as to the wording of such an agreement. The rule is written as if it expects an adjudication to have commenced, as there is a requirement on the parties to notify the adjudicator and for the adjudication to be suspended. However, there is nothing to prevent the parties pursuing mediation before adjudication but this may result in lengthening the time for resolving the dispute.

The adjudicator, if one is appointed, cannot act as mediator and the parties have seven days to agree on the mediator, otherwise CEDR will make an appointment if required by either party. No time is specified in which CEDR is to appoint but the rule requires that adjudication shall recommence if within 28 days of the agreement to mediate, a resolution has not been achieved. This also applies if either party abandons the mediation, and presumably the time period restarts from whatever day was reached before the suspension. Here again it is anticipated that adjudication will have been started prior to mediation, but where this is not so, one party may then refer the dispute to adjudication.

CIC Model Adjudication Procedure

The Model Adjudication Procedure known as the MAP was produced by a task force which represented a wide range of construction bodies, including those organisations which had or were to develop their own procedures, such as ICE and ORSA.

The model procedure comprises 35 brief rules and incorporates a model form of adjudicator's agreement. It was produced to comply with section 108 of the HGCRA. The model form lists its rules under the following headings:

- general principles
- application
- appointment of adjudicator
- conduct of adjudication
- the decision
- miscellaneous provisions
- definitions.

The general principles and application are broadly the same as those of the ICE and CEDR adjudication procedures. However, these provisions contain a clear statement regarding differences between the contract and the procedure and that the latter will prevail, except where the contract provides otherwise.

The appointment of the adjudicator is also dealt with in a similar way to other rules in that either party can refer a dispute to adjudication, but clause 9 is explicit in that the object is to secure the appointment and referral of the dispute within seven days of the notice of a dispute. This accords precisely with the requirements of section 108 of the HGCRA but of course the actual provisions must also achieve this object. Most adjudication procedures recognise that an adjudicator may be either named in the contract or appointed afterwards and this procedure is no exception. The principal difference lies in the actual timings. Where named in the contract the adjudicator is required to confirm availability within two days. Failing this, or if no adjudicator is named in the contract, the appointing body named in the contract, or the CIC if none is stated, must nominate an adjudicator within five days of the request. Of course, the parties may themselves agree the adjudicator but to ensure compliance with the Act a two-day period from the giving of the dispute notice is provided. There is also provision for a replacement adjudicator when required.

An objection by a party to the appointed adjudicator (rule 13) does not invalidate any decision reached. This is an interesting rule because it makes clear that although there is nothing to prevent an objection, it will be of no effect. The ICE and CEDR are not explicit on this matter although the latter did cover this point in an earlier draft of the procedure.

The conduct of the adjudication is covered in detail and generally in similar vein to other procedures, but the important date of referral is taken to be that on which the adjudicator receives the statement of case from the referring party.

In the adjudication the adjudicator is said to have complete discretion as to conduct and a list of actual powers is also provided. This list is not limiting in any way and perhaps serves more as an aide memoire than anything else. Under rule 18 requests and directions of the adjudicator must be complied with but if there is any failure in this regard there is no specified action that can be implemented.

Joinder of other parties is provided for in rule 22 but it requires the agreement of the adjudicator and the other parties before it can occur. Joinder is also provided for in the ICE Adjudication Procedure but not in the CEDR scheme. As agreement is required by all concerned under rule 22, it seems of little significance whether such a rule exists. It would only be significant if one or more parties had no choice but to accept the joinder of another.

The provisions relating to the decision are very similar to those contained within the ICE and no further comment is necessary. Liability of the adjudicator is covered in the miscellaneous provisions and clauses 33 and 34 are an attempt to exclude liability to the parties and to third parties. Rule 34 states:

'The adjudicator is appointed to determine the dispute or disputes between the Parties and his decision may not be relied upon by any third parties, to whom he shall owe no duty of care.'

It is doubtful that this will achieve its objective of giving immunity to the adjudicator in third party actions. This provision may not be known to the third party and the whole circumstances would need to be examined to see if a duty of care had been established against the backdrop of negligent mis-statement cases. The words contained in the second sentence of clause 7.2 of the ICE procedure, where the parties indemnify the adjudicator, are far better.

Adjudicator's agreement

The agreement follows a similar format to that of the other agreements discussed in this book. Each, of course, refers to their own specific adjudication procedure but otherwise there are distinct similarities and it is questionable why so many versions are necessary.

A confidentiality clause is contained in the agreement but only extends to the adjudicator and the parties. It may be wise to ensure this clause is amended to cover anyone else involved in the adjudication.

The immunity of the adjudicator contained in rule 33 is also provided for in the adjudicator's agreement to make sure the parties are aware of the adjudicator's liability and formally agree to the position.

CIC Adjudication Advice Service

In January 1998 at a meeting of the CIC adjudication task force it was decided to set up an advice service to assist construction professionals to adjust to the implementation of the statutory right to adjudication. A team of experts was identified to answer questions from the industry and guidance is circulated by means of the CIC Digest. Clearly, this initiative is worthwhile as it allows experience to be readily shared and the implementation of the adjudication part of the HGCRA to be monitored.

ORSA Adjudication Rules (now the TecSA Rules)

The Official Referees Solicitors Association (ORSA) produced its Procedural Rules for Adjudication in 1996 and was quickly off the mark in providing rules to meet the requirements of the HGCRA. The original version (1.1) was superseded by ORSA Adjudication Rules 1998 version 1.2 published in April 1998. These later rules were basically the same but amendments were made to rule 3 discussed later; otherwise the following discussions applies to both versions.

The rules can be incorporated into any contract and become operative when any party under the contract gives written notice of a dispute requiring adjudication. ORSA guidance notes suggest the following wording:

'Any dispute arising under this agreement shall in the first instance be referred to adjudication in accordance with the ORSA Adjudication Rules.'

If these rules or other properly drafted rules are not stated in the contract or incorporated therein, the terms of the Scheme for Construction Contracts would apply.

The rules make it clear that any number of disputes can be referred and this can occur regardless of whether arbitration of litigation has been started. Version 1.2 published in 1998 made two additions to rule 3 dealing with commencement. These rules require that:

'within 7 days from the date of notice, and provided that he is willing and able to act, any agreed Adjudicator ... nominated Adjudicator ... replacement Adjudicator ... shall give written notice of his acceptance of appointment to all parties.'

The date of confirmation becomes the date of the referral of the dispute.

These rules deal with the twin situations where the parties have agreed the adjudicator and where they have not. If already agreed, the adjudicator must confirm that he/she is able to start within seven days of referral. If not agreed the parties may apply to ORSA for an appointment of adjudicator to be made. The ORSA will endeavour to make the appointment to allow referral of the dispute within seven days from the notice requiring adjudication.

The ORSA has the power to replace the adjudicator on written notice if either party represents that the adjudicator is not performing properly. The circumstances that may give rise to this are wide and include lack of impartiality, incapacity and failing to act diligently or within the prescribed timescales. It is always a problem in seeking a replacement adjudicator that time will often be extended and to ensure the effect of this is minimised any direction and decision of a replaced adjudicator remains in force until reviewed.

Scope, purpose and conduct of adjudication

The rules do not set out the general principles on which adjudication is based but the broad provisions are not dissimilar to the other adjudication procedures discussed. The adjudication is confidential and the adjudicator is protected from claims except where bad faith is present. This immunity does not extend to third party actions and is in line with the CIC procedure rather than the extended immunity provided under the ICE procedure.

Under rule 11 the notice of adjudication determines the scope of the adjudication. The parties can agree to extend this and the adjudicator can also do this where it is essential to make the adjudication effective.

Speed and economy of the adjudication are important issues and the adjudicator will have this in mind in the choice of procedures. The adjudicator has wide discretion as to procedure and can open up and review certificates and other things issued under the contract. It is interesting that initially the adjudicator must assess the legal entitlements and only if practical constraints prevent this, for example time, can the adjudicator be more lax. If the decision cannot be based squarely on legal entitlements then it is to be a 'fair and reasonable view'. This accords with one view of adjudication that it should enable a non-legal view to prevail. However, the legal position will often be determined because where it is not, the decision, which is binding until finally determined by legal proceedings or arbitration, is more likely to be challenged.

Rule 19 lists a range of powers that the adjudicator can use. These do not limit the adjudicator's wide discretion. The list does not contain anything that is very surprising but three points are of note:

(1) the adjudicator can limit the length of any written or oral submission,
(2) specialist advice may be obtained but only one party need consent or request such help,
(3) the conduct of the adjudication is specifically referred to as 'inquisitorial'.

Point 1 is an important power, the proper use of which will facilitate speedy and economical dispute resolution.

Point 2 allows the adjudicator to obtain specialist help where appropriate and the parties can request that such specialist help be obtained. It is a decision for the adjudicator whether such help is sought but where it is the adjudicator's own initiative, at least one party must consent. Good practice suggests that the adjudicator should seek the consent of both parties before obtaining specialist help for which the parties have to pay.

Point 3 relates to the adjudicator using his initiative in ascertaining the facts and the law and this is generally what is required in an adjudication (section 108 of the HGCRA). The word 'inquisitorially' is an attempt to secure this by showing that it is a different process from that adopted in the courts, although arbitrators can now proceed in this way under section 34 of the Arbitration Act 1996.

The various adjudication procedures emphasise the wide powers of an adjudicator but the constraints are less obvious, even here. In the ORSA procedure, rule 21 sets out clearly some very important restrictions. Fairly common are those relating to written submissions from one party being made available to the other party, the adjudicator not acting when there is a conflict of interest and having no power to require one party to pay the legal costs of the other. However, item (i) which prevents advance payment or security of the adjudicator's fees and item (iii), refusal to allow a party to be represented, should be noted.

First, the adjudicator may not be content to act and provide a decision without either an advance payment or security for his/her fees. Where this is the case he should not consent to operate under the ORSA procedure. This provision also prevents the delivery of the decision being subject to full payment. It does not, however, cover the cost of specialists; advance payment of or security for these fees could be sought even under the ORSA rules. Second, the parties may if they choose be represented by anyone during the adjudication and the adjudicator cannot refuse that right. This again means that financially stronger parties may use 'high-powered' representatives to the disadvantage of the weaker party. The risk is always there that the lawyers will take over. Other procedures do not prevent such representation but rather tend to play it down. This is a difficult issue because a party should not be denied a proper opportunity to put their argument, even if that means getting someone else to do it. But one must be careful that adjudication does not simply become a complex lower tier of dispute resolution.

Decision

The adjudicator's decision is required within 28 days of the referral of the dispute or within 42 days if the referring party consents to a request for an extra 14 days from the adjudicator. The parties can agree any longer period so long as this is done after the dispute is referred. This is all in accordance with the HGCRA. The decision need not include any reasons but must be in writing and may include directions as to compound or simple interest.

Until the dispute is finally determined in arbitration or legal proceedings it will be binding on the parties and a decision must be implemented immediately. Summary enforcement of decisions through the courts is available even where other proceedings have been or are about to be commenced. Rule 28 states:

'No party shall be entitled to raise any right of set-off, counterclaim or abatement in connection with any enforcement proceedings'

in order to avoid summary judgment/interim payment applications being thwarted by defendant affidavits.

The adjudicator's decision can be challenged in arbitration, if an arbitration agreement exists or the parties so agree, or in legal proceedings. There is no stated time period in the ORSA rules during which an adjudicator's decision can be challenged but this may be governed by a provision in the principal contract. However, no application can be made to the courts until the adjudicator has made a decision and the party making such an application has complied with that decision or the adjudicator has refused to make a decision. The conduct of the adjudication itself may be challenged in the courts only where it involves the adjudicator acting in bad faith, want of jurisdiction, fraud or collusion.

If the adjudicator's decision is challenged the adjudicator cannot be involved in or be required to provide information in subsequent proceedings in litigation or arbitration.

Adjudicator's fees and expenses

So who pays? The parties share the adjudicator's fees unless the adjudicator directs otherwise, and pay their own costs. The parties are jointly responsible for the fees and expenses of the adjudicator except where an adjudication has been wrongly requested and here the party requesting the adjudication becomes solely responsible.

Fees are more fully dealt with in the adjudicator's agreement. They are included in rule 25 where an adjudicator's daily rate of £1000 maximum is specified.

Appendix 1

Australian Standard AS 2124–1992

46.2 Time for Disputing Superintendent's Direction
If the Superintendent –
(a) has given a direction (other than a decision under Clause 47.2) pursuant to the Contract; and
(b) has served a notice in writing on each party that if a party wishes to dispute the direction then that party is required to do so under Clause 47,
the direction shall not be disputed unless a notice of dispute in accordance with Clause 47.1 is given by one party to the other party and to the Superintendent within 56 days after the date of service on that party of the notice pursuant to Clause 46.2(b).

47. Dispute Resolution
47.1 Notice of Dispute
If a dispute between the Contractor and the Principal arises out of or in connection with the Contract, including a dispute concerning a direction given by the Super-intendent, then either party shall deliver by hand or send by certified mail to the other party and to the Superintendent a notice of dispute in writing adequately identifying and providing details of the dispute.

Notwithstanding the existence of a dispute, the Principal and the Contractor shall continue to perform the Contract, and subject to Clause 44, the Contractor shall continue with the work under the Contract and the Principal and the Contractor shall continue to comply with Clause 42.1.

A claim in tort, under statute or for restitution based on unjust enrichment or for rectification or frustration, may be included in an arbitration.

47.2 Further Steps Required Before Proceedings
Alternative 1
Within 14 days after service of a notice of dispute, the parties shall confer at least once, and at the option of either party and provided the Superintendent so agrees, in the presence of the Superintendent, to attempt to resolve the dispute and failing resolution of the dispute to explore and if possible agree on methods of resolving the dispute by other means. At any such conference each party shall be represented by a person having authority to agree to a resolution of the dispute.

In the event that the dispute cannot be so resolved or if at any time either party considers that the other party is not making reasonable efforts to resolve the dispute, either party may by notice in writing delivered by hand or sent by certified mail to the other party refer such dispute to arbitration or litigation.

Alternative 2
A party served with a notice of dispute may give a written response to the notice to the other party and the Superintendent within 28 days of the receipt of the notice.

Within 42 days of the service on the Superintendent of a notice of dispute or within 14 days of the receipt by the Superintendent of the written response, whichever is the earlier, the Superintendent shall give to each party the Superintendent's written decision on the dispute, together with reasons for the decision.

If either party is dissatisfied with the decision of the Superintendent, or if the Superintendent fails to give a written decision on the dispute within the time required under Clause 47.2 the parties shall, within 14 days of the date of receipt of the decision, or within 14 days of the date upon which the decision should have been given by the Superintendent confer at least once to attempt to resolve the dispute and failing resolution of the dispute to explore and if possible agree on methods of resolving the dispute by other means. At any such conference, each party shall be represented by a person having authority to agree to a resolution of the dispute.

In the event that the dispute cannot be so resolved or if at any time after the Superintendent has given a decision either party considers that the other party is not making reasonable efforts to resolve the dispute, either party may, by notice in writing delivered by hand or sent by certified mail to the other party, refer such dispute to arbitration or litigation.

47.3 Arbitration

Arbitration shall be effected by a single arbitrator who shall be nominated by the person named in the Annexure, or if no person is named, by the Chairperson for the time being of the Chapter of the Institute of Arbitrators Australia in the State or Territory named in the Annexure...

Unless the parties agree in writing, any person agreed upon by the parties to resolve the dispute pursuant to Clause 47.2 shall not be appointed as an arbitrator, nor may that person be called as a witness by either party in any proceedings.

Notwithstanding Clause 42.9 the arbitrator may award whatever interest the arbitrator considers reasonable.

If one party has overpaid the other, whether pursuant to a Superintendent's certificate or not and whether under a mistake of law or fact, the arbitrator may order repayment together with interest.

The Australian Standard AS 4300–1995

46.5 Time for Disputing Superintendent's Direction

If the Superintendent –

(a) has given a direction (other than a decision under Clause 47.2) pursuant to the Contract; and

(b) has served a notice in writing on each party that if a party wishes to dispute the direction, then that party is required to do so under Clause 47,

the direction shall not be disputed unless a notice of dispute in accordance with Clause 47.1 is given by one party to the other party and to the Superintendent within 28 days of the date of service on that party of the notice pursuant to Clause 46.5(b).

47. Dispute Resolution

47.1 Notice of Dispute

If a dispute or difference (hereafter called a 'dispute') between the Contractor and the Principal arises in connection with the Contract or the subject matter thereof, including a dispute concerning –

(a)   a direction given by the Superintendent; or
(b)   a claim –
   (i)   in tort;
   (ii)   under statute;
   (iii) for restitution based on unjust enrichment; or
   (iv)   for rectification or frustration,

then either party shall deliver by hand or send by certified mail to the other party and to the Superintendent a notice of dispute in writing adequately identifying and providing details of the dispute.

Notwithstanding the existence of a dispute, the Principal and the Contractor shall continue to perform the Contract and, subject to Clause 44, the Contractor shall continue with the work under the Contract and the Principal and the Contractor shall continue to comply with Clause 42.1.

47.2   Further Steps Required Before Proceedings

Alternative 1

Within 14 days of service of a notice of dispute, the parties shall confer at least once to attempt to resolve the dispute or to agree on methods of resolving the dispute by other means. At any such conference each party shall be represented by a person having authority to agree to a resolution of the dispute.

If the dispute has not been resolved within 28 days of service of the notice of dispute, that dispute shall be and is hereby referred to arbitration.

Alternative 2

A party served with a notice of dispute may give a written response to the notice to the other party and the Superintendent within 28 days of the receipt of the notice.

Within 42 days of service on the Superintendent of a notice of dispute or within 14 days of the receipt by the Superintendent of the written response, whichever is the earlier, the Superintendent shall give to each party the Superintendent's written decision on the dispute, together with reasons for the decision.

If either party is dissatisfied with the decision of the Superintendent or if the Superintendent fails to give a written decision on the dispute within the time required under this Clause 47.2, the parties shall, within 14 days of the date of receipt of the decision or within 14 days of the date upon which the decision should have been given by the Superintendent, confer at least once to attempt to resolve the dispute or to agree on methods of resolving the dispute by other means. At any such conference, each party shall be represented by a person having authority to agree to a resolution of the dispute.

If the dispute has not been resolved within 28 days of the date the Superintendent has given a decision or should have given a decision, that dispute shall be and is hereby referred to arbitration.

Appendix 2

Extracts from Australian Department of Defence contracts

The following is an extract from a suite of contracts introduced by the Australian Department of Defence for use in work done for the department including similar provisions to those contained in the Australian Standards AS 2124–1992 and AS 4300–1995.

45. Settlement of Disputes
45.1 Method of Resolution
All disputes or differences between the Principal and the Contractor or between either the Principal or the Contractor and the Superintendent at any time must be determined as follows:
(a) if the dispute or difference concerns a direction of the Superintendent under one of the clauses or sub-clauses referred to in Annexure One, the dispute or difference must be determined in accordance with the procedure under sub-clauses 45.4 to 45.11; or
(b) in the case of any other dispute or difference as to:
  (i) the construction of this Contract, or
  (ii) any fact, matter or thing of whatsoever nature arising out of or in connection with the Contract or the work performed by the Contractor, or
  (iii) any decision of an Adjudicator under sub-clause 45.9 where notice has been given as required by sub-clause 45.12,
  the dispute or difference must be determined by arbitration.

45.2 Adjudicator
The Contractor shall within seven (7) days of the date of acceptance of tender give notice in writing to the Superintendent nominating:
(a) one of the persons referred to in Annexure One as the person to act as the Adjudicator for the purposes of this clause; and
(b) the order of preference in which the remaining persons referred to in Annexure One shall act as the Adjudicator pursuant to sub-clause 45.2.
If the Contractor fails to give this notice within the time required the Superintendent shall give notice in writing to the parties nominating the matters referred to in sub-paragraphs (a) and (b).

45.3 Substituted Adjudicator
If for any reason whatsoever, the person chosen by the Contractor or the Superintendent (as the case may be) to act as the Adjudicator is unavailable to act as the Adjudicator under this clause in respect of a dispute or difference then the person who is next available in the order of preference nominated by the Contractor or the Superintendent (as the case may be) pursuant to sub-clause 45.2 will act as the Adjudicator.

45.4 Adjudication Dispute or Difference

In the event of any dispute or difference arising between the Principal and the Contractor or between either the Contractor or the Principal and the Superintendent at any time as to any direction under one of the clauses or sub-clauses referred to in Annexure One, then either party, if it wishes to have such direction reviewed, shall give to the other party and to the Superintendent notice in writing, which notice shall comply with the requirements of sub-clause 45.5. Such notice shall be given within fourteen (14) days of the dispute or difference arising.

The party that gives a notice under this sub-clause 45.4 shall hereafter be referred to as 'the dissatisfied party'.

The review procedure in sub-clauses 45.4 to 45.13 does not apply to any direction of a Superintendent's Representative. In the event either party is dissatisfied with any direction of a Superintendent's Representative it shall proceed in accordance with the procedure prescribed in Clause 49.

Unless a notice is given in accordance with the first paragraph of this sub-clause within the time prescribed, the Superintendent's direction shall be final and binding upon the parties.

45.5 Requirements of Adjudication Notice

The notice in writing given by the dissatisfied party pursuant to sub-clause 45.4 shall:
(a) specify that it is a notice given under sub-clause 45.4;
(b) identify the clause or sub-clause under which direction the subject of the dispute or difference was given; and
(c) be accompanied with adequate particulars and any relevant written material which identifies the matters the subject of the dispute or difference in relation to the direction.

If the notice is not given within the fourteen (14) days stipulated in sub-clause 45.4, the failure to give the notice shall be an absolute bar on the giving of such notice at any time thereafter and in addition such failure shall operate as a complete and unconditional waiver by each party to object in any way, at any time and for any reason to the matters the subject of the dispute or difference in relation to the direction.

45.6 Notification of Adjudicator

After receiving the notice under sub-clause 45.4, the Superintendent shall:
(a) within five (5) days of receiving the notice inform the Adjudicator of the existence of the relevant dispute and request confirmation from the Adjudicator within a further five (5) days that the Adjudicator is available to act in accordance with the procedure set down in this Clause;
(b) if the Adjudicator is not available, advise the person next in the order of preference nominated by the Contractor or the Superintendent (as the case may be) pursuant to sub-clause 45.2 of the existence of the dispute and seek confirmation that such person is available to act as the adjudicator; and
(c) if no person named in Annexure One is available to act as the Adjudicator, request the President or acting President for the time being of the Institute of Arbitrators Australia to nominate a person to act as the Adjudicator.

After the Adjudicator is chosen in accordance with the above procedure the Superintendent shall despatch the notice provided by the dissatisfied party pursuant to sub-clause 45.4 to the Adjudicator.

The Superintendent shall thereafter have no other contact with the Adjudicator prior to the meeting referred to in sub-clause 45.7 and neither party shall have any contact with the Adjudicator except for the purpose of agreeing administrative details with regard to the time and location of the meeting referred to in sub-clause 45.7.

45.7 Meeting with Adjudicator

Within five (5) days after the provision of the notice under sub-clause 45.4 by the Superintendent to the Adjudicator the parties and the Superintendent shall meet with the Adjudicator to agree the procedure to be adopted in resolving the dispute or difference, and failing agreement the procedure shall be determined by the Adjudicator.

45.8 Powers of Adjudicator

The Adjudicator shall have the power:
(a) to open up, review and revise any direction of the Superintendent which has been referred to him as if no such direction had been given and substitute his exercise of any discretion for that of the Superintendent;
(b) to proceed to the resolution of the dispute or difference in such manner and subject to such rules as the Adjudicator and the parties agree or failing agreement as the Adjudicator in his absolute discretion determines is suitable for the nature of the dispute or difference and the Adjudicator shall not be bound by rules of evidence nor shall the parties have any right of legal representation; and
(c) to engage and consult with any advisors, legal or technical, as he may see fit.

45.9 Adjudicator to Give Decision

Within seven (7) days after the conclusion of the dispute resolution procedure agreed or determined under sub-clause 45.7, the Adjudicator shall notify the parties in writing of his decision and shall not be required to give reasons for that decision.

45.10 Adjudicator to Act as Expert

In making his decision the Adjudicator shall act as an expert and not as an Arbitrator.

45.11 Adjudicator's Costs

The parties shall pay the Adjudicator's costs (including the costs of engaging and consulting advisors pursuant to sub-paragraph (c) of sub-clause 45.8) equally.

45.12 Dissatisfaction with Adjudicator's Decision

If either party is dissatisfied with the decision of the Adjudicator pursuant to sub-clause 45.9 it shall give notice in writing to the other party within fourteen (14) days from the date of receipt of the Adjudicator's decision requiring the dispute or difference to be referred for discussion pursuant to sub-clause 45.13.

If such notice is not given within the said fourteen (14) days, the decision of the Adjudicator shall thereafter be final and binding and not subject to review in any court or by arbitration.

45.13 Executive Negotiations

If either:
(a) a notice is given pursuant to sub-clause 45.12; or
(b) a dispute or difference arises between the Principal and the Contractor or between either the Contractor or the Principal and the Superintendent as to any

matter other than one to which sub-clause 45.4 applies and either party gives notice in writing to the other party adequately identifying the matters the subject of the dispute or difference together with detailed particulars thereof, the persons described in Annexure One (''the Parties' representatives'') shall use their best endeavours to:

(c) resolve the dispute or difference, and for this purpose they shall undertake such investigations, hold such meetings and conduct such informal hearings as is thought necessary; or

(d) if the dispute or difference cannot be resolved, agree upon a process for resolving the whole or part of the dispute or difference by means other than litigation or arbitration.

45.14 Arbitration Agreement

After the expiration of thirty (30) days from the giving of a notice under sub-clause 45.12 or sub-clause 45.13(b) (as the case may be), all disputes and differences not resolved by the procedures set out in the preceding sub-clauses shall be referred to arbitration. The dispute or difference shall be identified together with detailed particulars in a notice in writing served by the party who served the notice under sub-clause 45.12 or sub-clause 45.13(b) (as the case may be) upon the other party. This notice shall be given after the expiration of thirty (30) days from the giving of the notice under sub-clause 45.12 or sub-clause 45.13(b) (as the case may be). The parties agree that they may each be represented by a legal practitioner in any arbitration proceedings conducted pursuant to this clause 45.

45.15 Identity of Arbitrator or Arbitrators

Any arbitration under this clause shall be effected:

(a) By either a single arbitrator or by two arbitrators agreed upon in writing between the Principal and the Contractor; or

(b) failing such agreement within ten (10) days after receipt by the other party of the notice in writing given under sub-clause 45.14 by the one party then by an arbitrator, to be selected by the President or acting President for the time being of the Institute of Arbitrators Australia; or

(c) if he fails to select an arbitrator then an arbitrator shall be appointed pursuant to the laws of the State or Territory in which the Works are situated.

45.16 Arbitrator's Powers

The Arbitrator shall have the power from time to time:

(a) to make any order in regard to the provision of further security for the costs of the arbitration proceedings;

(b) to direct in what manner any security for costs of the arbitration proceedings shall be applied;

(c) to open up, review and revise any direction of the Superintendent or decision of the Adjudicator as if no such direction or decision had been given and substitute the exercise of his discretion for that of the Superintendent or the Adjudicator; and

(d) subject to sub-clause 45.1, allow either party at any stage of the proceedings to raise by way of further claim, set-off, defence or cross-claim and subject to any condition as to costs or otherwise that may be imposed by the Arbitrator any dispute or difference whatever relating to the construction of the Contract or as to any fact, matter or thing of whatsoever nature arising out of or in connection with the Contract or the work under the Contract.

45.17 Arbitration Costs

The costs of a submission, reference and award under this clause and the apportionment thereof shall be at the discretion of the Arbitrator.

45.18 Proceeding with the Works

Notwithstanding the existence of a dispute or difference:

(a) the Contractor shall if the Works (including the making good of any defects under clause 37) have not been completed at all times (subject as otherwise may be provided for in the Contract) proceed without delay to continue to perform and execute the Works and in so doing shall comply with all instructions of the Superintendent; and

(b) the Principal shall continue to comply with its obligations under the Contract.

Appendix 3

Adjudicator Nominating Bodies

The Department of the Environment Transport and the Regions has compiled a list of adjudicator nominating bodies from those who indicated a willingness to offer this service. The following organisations are currently listed for England and Wales and a separate list is maintained by the Scottish Office for Scotland.

Academy of Construction Adjudicators
Centre for Dispute Resolution
Chartered Institute of Arbitrators
Chartered Institute of Building
Confederation of Construction Specialists
Construction Industry Council
Construction Confederation
Institution of Chemical Engineers
Institution of Electrical Engineers
Institution of Civil Engineers
Official Referees Bar Association (now TecBA)
Official Referees Solicitors Association (now TecSA)
Royal Institute of British Architects
Royal Institution of Chartered Surveyors
3As
Institution of Mechanical Engineers

Appendix 4
The Secretary of State for Transport and Severn River Crossing plc Concession Agreement

Schedule 5   *Disputes Resolution Procedure*

General

1. If a Dispute arises, whether before or after the Concession Commencement Date and whether before or after repudiation or other termination of this Concession Agreement then either party may refer the Dispute to the decision in the first place of the Panel acting as independent experts but not as arbitrators.

2. Any unanimous decision of the Panel (including for the avoidance of doubt, a quorate Panel) shall be final and binding upon the parties but otherwise a decision of the Panel shall be final and binding upon the parties unless and until the Dispute has either been settled or referred to arbitration as hereinafter provided and an arbitral award has been made. It shall be a condition precedent to the commencement of any action at law that, in relation to the subject matter of such action, there shall have been either:

   2.1   a unanimous decision of the Panel; or

   2.2   an arbitral award; or

   2.3   a settlement agreement between the parties.

3. Unless both parties otherwise agree in writing any representations or concessions made by either party in or in connection with proceedings before the Panel or any representations, concessions or agreements (other than a settlement agreement) made by either party in the course of discussions pursuant to Schedule 5, paragraph 11 between the chief executive officer of the Concessionaire and the official nominated for that purpose by the Secretary of State shall be without prejudice and shall not be raised by either party in any subsequent arbitration or other legal proceedings.

4. Unless this Concession Agreement has already been repudiated or terminated, the Concessionaire shall in every case continue to proceed with the Works with all due diligence regardless of the nature of the Dispute and the parties shall give effect forthwith to every decision of the Government's Agent or the Government's Representative or the Panel except and to the extent that the same shall have been revised by settlement agreement or arbitral award.

5. In this Schedule 5, the expression 'the Panel' may be construed to mean either the Financial Panel or the Technical Panel whichever shall be appropriate in the context. The expression 'the chairman' shall similarly be construed to mean either the chairman of the Financial Panel or the chairman of the Technical Panel.

The Panel

6.1 The Panel shall conduct the reference and make its decision in accordance with the Panel Rules.

6.2 Save as otherwise expressly provided the Panel shall have power to open up, review and revised any certificate, opinion, instruction, notice, statement of objection, determination or decision of any person given or made pursuant to this Concession Agreement.

7.1 In the event that any member of the Panel shall become unable or unwilling to act either at all or on such occasions or for such periods as to render it necessary or expedient, in the opinion of the chairman, for a replacement to be appointed, the parties shall agree and appoint such replacement. In default of such agreement within 28 days of notification to the parties by the chairman of the need for a replacement, the chairman shall recommend a candidate having ascertained his suitability and willingness to act. Such person shall be appointed to the Panel unless both parties agree otherwise in which case the chairman shall recommend other candidates until the approval of at least one party to an appointment is obtained whereupon the candidate so approved shall be appointed to the Panel.

7.2 Where there is a Dispute which has been referred to the Panel then, notwithstanding Schedule 5, paragraph 7.1, in the event that any member of the Panel shall be or become unable or unwilling to act in relation to such Dispute and in the absence of agreement between the parties as to the appointment of an ad hoc replacement or of an appointment pursuant to Schedule 5, paragraph 7.1 within 7 days of notification by the chairman of the need for a replacement, either party may apply to the relevant nominating authority referred to in Schedule 5, Appendix 2 to appoint a replacement to sit on the Panel for the purpose of deciding the said Dispute but for no other purpose.

8. If an appointment is made pursuant to Schedule 5, paragraph 7 and if the period between the date of appointment and the date that a decision ought to be made by the Panel comprising such replacement pursuant to Panel Rule 6.1 is less than 21 days, then the parties shall be deemed to have agreed that the time limit under Panel Rule 6.1 shall be extended by such number of days as may be necessary to give the replacement 21 days from the date of his appointment to consider such reference.

9. Subject to Panel Rule 7.1:

9.1 the parties shall bear equally the costs of and incidental to the engagement of the members of the Financial Panel and the Technical Panel save and except those charges and expenses described in Schedule 5, paragraph 9.2; and

9.2 charges and expenses attributable to disputes resolution under the Construction Contract to which the Secretary of State does not become a party under Panel Rule 8.1.3 and the costs of the Technical Panel after the issue of the maintenance certificate under the Construction Contract when there is no unresolved Dispute before the Technical Panel shall as between the Secretary of State and the Concessionaire be borne by the Concessionaire; and

the Secretary of State shall pay all such sums properly due and shall issue an invoice to the Concessionaire in accordance with Clause 41 [Payment] for one half of payments to be borne in accordance with Schedule 5, paragraph 9.1 and for the whole of payments to be borne in accordance with Schedule 5, paragraph 9.2.

10.1 The Technical Panel shall be maintained from the date of execution of this Concession Agreement until 28 days after the date of issue of the final certificate for the purposes of the Construction Contract or until the date of determination of a reference commenced prior to that date whichever shall be the later. Any Dispute which would have been referred to the Technical Panel arising thereafter shall be referred to arbitration.

10.2 The Financial Panel shall be maintained from the date of execution of this Concession Agreement until such time as the parties may agree that it should be stood down. If a Dispute should arise after the Financial Panel has been stood down, the parties shall reconstitute the Financial Panel including so far as practicable the members of the Financial Panel at the time it was stood down. In the event that the parties cannot agree the appointment of any or any further members of the Financial Panel, the requisite appointments shall be made by the nominating authority referred in Schedule 5, Appendix 2, Part I.

Procedure for amicable settlement

11. In the event that either party wishes to contest a non-unanimous decision of the Panel or in the event that the Panel shall have failed to make a decision in accordance with Panel Rule 6.1, the Dispute shall be referred by notice in writing to the chief executive officer of the Concessionaire and the official of the Department of Transport nominated for that purpose by the Secretary of State who shall meet and endeavour to resolve the issues between them. The joint and unanimous decision of the said chief executive officer and the said official shall be binding upon the parties but if they are unable to agree within 28 days of the reference to them then either party may require the Dispute to be referred to arbitration.

Arbitration

12. Subject always to Schedule 5, paragraph 11, either party may require a Dispute to be referred to arbitration in the event that the Panel shall have failed to make a decision unanimously or in accordance with Schedule 5, paragraph 10.1 or Panel Rule 6.1.

13. In the case of Disputes touching or concerning the Works, a reference to arbitration may proceed prior to the date of issue of the Permit to Use or the date upon which the same ought to have been issued, provided that the obligations of the Secretary of State, the Government's Agent, the Government's Representative and the Concessionaire shall not be altered by reason of the arbitration being conducted during the progress of the Works. Save as aforesaid no steps shall be taken in a reference of any such Dispute to arbitration until the issue of the Permit to Use unless the parties shall otherwise agree in writing.

14. Disputes which are to be the subject of arbitration shall be referred to one arbitrator to be agreed between the parties and in default of agreement to be appointed in the case of Disputes falling within Panel Rules 1.3.1 or 1.3.2 by the President or Vice-President for the time being of the Chartered Institute of Arbitrators and references of such Disputes shall be conducted in accordance with the said institute's arbitration rules or any amendment or modification thereof in force and in the case of other Disputes by the President for the time being of the Institution of Civil Engineers or in his absence the Vice President for the time being of the said institution and references of such other Disputes shall be conducted in accordance with the Institution of Civil Engineers'

Arbitration Procedure (1983) or any amendment or modification thereof in force. The place of arbitration shall be London. Each such reference shall be in writing, specifying the matter or issue in dispute, and shall state that it is made pursuant to the relevant Clause and Schedule of this Concession Agreement.

15. In the event that a Construction dispute referable to arbitration raises issues which are substantially the same as or connected with issues raised in a Dispute which has been referred to an arbitrator, the Secretary of State and the Concessionaire agree that:

15.1 the Concessionaire and/or the Contractor may refer the Construction Dispute to the arbitrator to whom the Dispute has been referred on the basis that such arbitrator shall have the powers described in Schedule 5, paragraph 15.2; and

15.2 such arbitrator shall have power to make such directions and all necessary awards in the same way as if the procedure of the High Court as to joining third parties was available to the parties and to him.

16. Save as expressly otherwise provided, the arbitrator shall have full power to open up, review and revise any decision, opinion, instruction, notice, statement of objection, determination or certificate of the Government's Agent or the Government's Representative related to the Dispute and any non-unanimous decision of the Panel and to order the rectification of this Concession Agreement and of any agreement made between the parties pursuant thereto subject to any rule of law which would restrict this power.

17. A past or present member of the Panel shall not be eligible for appointment as an arbitrator unless the parties otherwise agree in writing.

Schedule 5 Appendix 1   Panel Rules

1. *Commencement*

1.1 Either party may commence a reference under these rules by serving a notice upon either the chairman of the Technical Panel or the chairman of the Financial Panel in accordance with the provisions of Panel Rule 1.3.

1.2 A notice served under Panel Rule 1.1 shall include:

1.2.1 a concise summary of the nature and background of the Dispute and the issues arising; and

1.2.2 a statement of the relief claimed;

1.2.3 in respect of Disputes before the Technical Panel, a reference to all of the formal monthly reports submitted pursuant to Clause 10.10.1 [Design and Construction] of this Concession Agreement in which the subject matter of the Dispute was raised;

1.2.4 a statement of any matters which the parties have already agreed in relation to the procedure for determination of the Dispute;

1.2.5 copies of all documents which have an important and direct bearing on the issues and on which the claimant intends to rely (or a list of such documents if they are already in the possession of the recipient of the notice).

1.3 Unless the parties agree in writing to refer a Dispute to the Technical Panel or the chairman of the Financial Panel shall otherwise decide pursuant to Panel Rule 1.6, the claimant shall refer a Dispute to the Financial Panel if it arises out of or in connection with:

1.3.1 Clauses 1.8 [Interpretation], 4.1 [Provision of Guarantees], 5 [Financial Terms and Documentation], 6 [The Effect of the Bill, etc], 24 [Legislation], 26 [Insurance Obligations], 28 [Intellectual Property and Confidentiality], 29 [Tax], 30 [Force Majeure], 31 [Events of Default], 32 [Termination by Reason of Default]; or 38 [Custody of Financial Model] and/or

1.3.2 the adjustment of the ACRR, the RCRR, the Tolls or the Toll structure or any alleged entitlement to effect such an adjustment.

1.4 Unless the parties agree in writing to refer a Dispute to the Financial Panel or the chairman of the Financial Panel shall otherwise decide pursuant to Panel Rule 1.6 the claimant shall refer all Disputes not falling within Panel Rules 1.3.1 or 1.3.2 to the Technical Panel.

1.5 If the outcome of a Dispute referable to the Financial Panel may be affected to any material degree by the outcome of an unresolved Dispute which has been referred to the Technical Panel, no steps shall be taken in a reference of the former Dispute until the latter Dispute has been finally determined unless the parties shall otherwise agree or the Financial Panel shall otherwise order.

1.6 In the event of disagreement between the parties as to whether a Dispute or whether any particular issues in a Dispute should be referred to the Financial Panel or the Technical Panel, or as to the sequence in which Disputes or issues in a Dispute should be decided, the parties shall make representations to the chairman of the Financial Panel who, after consultation with the chairman of the Technical Panel, shall thereupon decide the appropriate forum or sequence as the case may be and, if a Dispute has already been the subject of a reference, the date to be deemed as the date of reference of the Dispute for the purposes of Panel Rule 6.1.

1.7 The Financial Panel and the Technical Panel shall each be bound by the decisions of the other so far as they may be material to a Dispute before them save to the extent that the same shall have been revised by settlement agreement or arbitral award.

2. *Quorum*

2.1 Unless the parties agree otherwise in writing, the quorum for all proceedings of:

2.1.1 the Technical Panel shall be all of its members present in person;

2.1.2 the Financial Panel shall be three of its members including the chairman present in person.

2.2 If the chairman of the Technical Panel forms the opinion that the nature, size or complexity of a Dispute is such that it would be equally or more appropriate for the Technical Panel to consist of three members, he shall recommend to the parties that they consent to the quorum for the purposes of determination of such Dispute being reduced accordingly. Neither party shall unreasonably withhold or delay its consent to such a recommendation.

3. *Procedure*

3.1 The Panel shall have the widest discretion permitted by law to determine its procedure (including without limitation the delegation of the power to make procedural rulings to its chairman) and to ensure the just, expeditious and economical determination of the Dispute after such investigation as the Panel may think fit provided that the Panel shall adopt all and any procedures agreed by the parties to be appropriate for determination of a Dispute.

3.2 Without prejudice to the generality of Panel Rule 3.1, the chairman of the Panel shall decide whether or not to convene a hearing or otherwise take oral evidence or proceed to determine the Dispute on a documents-only basis subject to the right of each party to make representations to the chairman of the Panel in relation to such matters.

3.3 The chairman of the Panel shall fix the date, time and place of any meetings, hearings, or inspections which the Panel deems appropriate, and shall give the parties and other members of the Panel reasonable notice thereof.

3.4 The chairman of the Panel may in advance of any hearing submit to the parties a list of questions which it wishes them to treat with special attention.

3.5 All meetings, hearings or inspections shall be in private unless the parties agree otherwise.

3.6 Each party may appoint representatives to appear on its behalf at a hearing, subject to such proof of authority as the Panel may require.

4. *Witnesses*

4.1 Before any hearing, the chairman of the Panel may require a party to give notice of the identity and qualifications of witnesses it wishes to call, and may require the parties to exchange statements of evidence to be given by the witnesses a specified time in advance of the hearing.

4.2 The Panel may allow, refuse or limit the appearance of witnesses, whether witnesses of fact or expert witnesses.

4.3 The Panel may commission expert evidence to be prepared and adduced by a witness independent of the parties. Unless otherwise agreed by the parties not more than two such witnesses may be called.

4.4 Any witness who gives oral evidence at a hearing may be questioned by each of the parties or their representatives under the control of the Panel. The Panel may put questions at any stage of the examination of a witness.

4.5 The Panel may allow the evidence of a witness to be presented in written form either as a signed statement or by a duly sworn affidavit. Any party may make representations that such a witness should attend for oral examination at a hearing. If the Panel so orders, and if the witness thereafter fails to attend, the Panel may place such weight on the written evidence as it thinks fit, or exclude it altogether.

5. *Powers of the Panel*

5.1 Without prejudice to Panel Rule 1.3 and to any powers which may be given to the Panel elsewhere in these Rules or in this Concession Agreement, the Panel shall have power:

(a) to examine any witness or conduct an inspection of any property or thing relevant to the Dispute in the absence of any or any other representative of the parties or any other person;

(b) at any time to permit any party to amend any submissions;

(c) to continue with the reference in default of appearance or of any act by any of the parties in like manner as a judge of the High Court might continue with proceedings in that court where a party fails to comply with an order of that court or a requirement of rules of court (including, for the avoidance of doubt and without limitation, power to strike out any claim, defence, counterclaim or other submission and to make any decision consequent upon any such striking out), in the event a party fails within the time specified in these Rules or in any order to do any act required by these Rules or to comply with any order;

(d) to order a party to produce to the other party and to the Panel for inspection, and to supply copies of, any documents in that party's possession, custody or power, which, in the event of dispute, the Panel determines to be relevant;

(e) to order a party to answer interrogatories on the application of the other party;

(f) to order the inspection, preservation, storage, interim custody, sale or other disposal of any property or thing relevant to the Dispute under the control of any party;

(g) to make orders authorising any sample to be taken, or any observation to be made, or experiment to be tried which may, in the Panel's discretion, be necessary or expedient for the purpose of obtaining full information or evidence;

(h) to require the parties to provide a written statement of their respective cases in relation to particular issues, to provide a written answer thereto and to give reasons for any disagreement.

5.2 For the avoidance of doubt, nothing in these Rules shall be taken as conferring power upon the Panel to order a party or a representative of a party to give evidence (whether in person or by way of documentary or similar evidence) which could not be ordered if the proceedings were before the High Court.

6. *Decisions*

6.1 The Panel shall make its decision within 28 days of the date of reference of a Dispute or within such other period as the parties may agree in writing.

6.2 The Panel shall make its decision in writing. Unless the parties otherwise agree, reasons for the decision and any reasons for dissent shall be given. The decision shall be dated and shall be signed or otherwise acknowledged in writing by all members of the Panel.

6.3 If a Panel member refuses or fails to sign or acknowledge the decision, the signatures of the majority shall be sufficient provided that the reason for the omitted signature is stated. Such a decision shall stand as a non-unanimous decision.

6.4 The Panel may allow interest on any sum which is the subject of a decision at such rates as the Panel determines to be appropriate and shall have the same (but no greater) powers in this respect as a judge of the High Court.

6.5 The Panel may make separate final decisions on different issues at different times.

6.6 Subject to Schedule 5, paragraph 2 decisions shall be final and binding on the parties as from the date upon which they are made.

7. *Costs*

7.1 The Panel shall have power to make a decision in respect of liability for all or part of the costs directly attributable to the reference by way of the charges and expenses of the members of the Panel and costs of an administrative nature (such as the hire of rooms) if the Panel considers that anything has been done, or that any omission has been made, unreasonably or improperly by or on behalf of a party, and to determine or assess the amount of those costs.

7.2 For the avoidance of doubt, the Panel shall not in any event have power to make a decision in respect of liability for the legal or other costs incurred by a party in preparing and presenting its case to the Panel. Such costs shall be borne by the party incurring them.

8. *Disputes between Concessionaire and Contractor*
8.1 In the event of a Construction Dispute being referred to the panel pursuant to the Construction Contract:

  8.1.1 the Concessionaire shall forthwith inform the Government's Agent and forward copies of all notices served; and

  8.1.2 the Concessionaire shall thereafter keep the Government's Agent fully informed as to progress of such reference and shall in particular but without limitation forward copies of all documents which are prepared by or come into the possession of the Concessionaire in connection with such reference and give notice to the Government's Agent of all hearings, meetings and inspections convened by the Panel; and

  8.1.3 if the Secretary of State in his sole discretion considers that the issues in such reference are or are potentially relevant to the rights and obligations of or issues between the parties to the Concession Agreement, he may by notice served on the Concessionaire, the Contractor and all members of the Panel, become a party to such reference and to have such rights, obligations or issues determined as a Dispute.

8.2 A notice served under Panel Rule 8.1.3 shall include:

  8.2.2 a concise summary of the rights and obligations of or the issues between the Secretary of State and the Concessionaire, identifying the issues as between the Concessionaire and the Contractor in the Construction Dispute before the Panel which are or are potentially relevant thereto; and

  8.2.3 a statement of relief claimed (whether by way of declaration or otherwise); and

  8.2.4 the like details and documents required by Panel Rules 1.2.3, 1.2.4 and 1.2.5.

8.3 In the event of service of a notice under Panel Rule 8.1.3:

  8.3.1 if the period between the date of service of such notice and the date for issue of a decision in the reference of the Construction Dispute is less than 21 days, all parties to the reference shall be deemed to have agreed that the time limit for issue of a decision in the tripartite reference shall be extended to expire 21 days from the date of service of such notice; and

  8.3.2 the Contractor shall be permitted to address the Panel on any point of construction of the Concession Agreement; and

  8.3.3 the decision of the Panel as between the Secretary of State and the Concessionaire shall, as between the Concessionaire and the Contractor, bind the Contractor as if the Contractor were a party to the Concession Agreement; and

  8.3.4 the charges and expenses attributable to the reference shall be borne equally between the Secretary of State, the Concessionaire and the Contractor.

9. *Contractor's Rights of Joinder*
9.1 In the event of a Dispute being referred to the Panel and if the issues in such reference are such that there is a material divergence between the interest of the Concessionaire and of the Contractor insofar as they depend on the decision of

the Panel, the Contractor may by notice served on the Secretary of State become a party to such reference and to have all relevant rights, obligations and issues under the Construction Contract determined as a Construction Dispute.

9.2 A notice served under Panel Rule 9.1 shall include:

9.2.1 a concise summary of the rights and obligations of or the issues between the Contractor and the Concessionaire, identifying the issues as between the Concessionaire and the Secretary of State in the Dispute before the Panel which are or are potentially relevant thereto; and

9.2.2 a statement of relief claimed (whether by way of declaration or otherwise); and

9.2.3 the like details and documents required by Panel Rules 1.2.3, 1.2.4 and 1.2.5.

9.3 In the event of service of a notice under Panel Rule 9.1:

9.3.1 if the period between the date of service of such notice and the date for issue of a decision in the reference of the Dispute is less than 21 days, all parties to the reference shall be deemed to have agreed that the time limit for issue of a decision in the tripartite reference shall be extended to expire 21 days from the date of service of such notice; and

9.3.2 Panel Rules 8.3.2, 8.3.3 and 8.3.4 shall apply.

10. *Exclusion of Liability*

Neither the chairman nor any member of the Panel shall be liable to any party for any act or omission in connection with any reference save that the chairman and any member of the Panel may be liable for the consequences of conscious and deliberate wrongdoing.

11. *Notices*

11.1 Unless otherwise ordered by the Panel, or agreed between the parties, all notices required by these Rules shall be in writing. A notice under Rule 1.1 shall be served by first class post or delivered by hand. All other notices and written communications shall be sent by first class post, fax or telex or delivered by hand.

11.2 Unless the intended recipient proves otherwise:

11.2.1 documents sent by first class post shall be deemed to have been received two working days after posting;

11.2.2 faxes or telexes shall be deemed to have been received at the time transmission ceases;

11.2.3 by hand deliveries shall be deemed to have been received at the time of delivery to the address stated on their face.

References in these Rules to receipt of documents shall be construed accordingly.

11.3 Notices shall be effective from the time of receipt. Periods of time measured with reference to the giving, sending, or serving of a document shall be measured with reference to the time that document is received.

11.4 Unless otherwise ordered by the Panel or agreed between the parties, all notices and other documents received on a day which is not a working day or after 6.00 p.m. on any working day shall be deemed to have been received on the following working day.

11.5 In every case in which a notice is sent to the chairman a copy thereof shall be sent to all other members of the Panel and to the other party or parties.

Schedule 5 Appendix 1 Part I   Composition of the financial panel

1. The Financial Panel shall consist of 5 members including the chairman.
2. The Secretary of State and the Concessionaire shall endeavour in good faith and with due expedition to agree the identity of persons willing and suitable to act as the initial members of the Financial Panel, and of one such person to act as its chairman, and of willing and suitable replacements whenever the need therefor arises.
3. In the event that the Secretary of State and the Concessionaire are unable to agree any or all of the initial members of the Financial Panel within 60 days of execution of this Concession Agreement, either party may apply to the nominating authority identified below to appoint such person or persons. Such party shall notify the nominating authority of the identity of those persons who have been considered by the parties under paragraph 2 above and who have been thought by either party not to be suitable. In the absence of agreement to the contrary of both parties the nominating authority shall not nominate any person who shall have been so identified.
4. The nominating authority for the purpose of appointing replacement members of the Financial Panel shall be the Director, Finance and Industry (or in his absence, the Head of the Industrial Finance Division) of the Bank of England.
5. The nominating authority may, if in his sole discretion he thinks fit, appoint fellow officers of the Bank of England to act as members of the Financial Panel (save that, for the avoidance of doubt, the nominating authority may not nominate himself or the then current alternate nominating authority as identified in paragraph 4 above).
6. With a view to facilitating the nominating authority's task of identifying suitable candidates for appointment, either party may forward to the nominating authority any relevant description which has been agreed by the parties of the likely background and experience of a suitable appointee and/or of the likely issues which the appointee as a member of the Financial Panel would be required to decide. The nominating authority may have regard to such description but shall not be bound by its terms.
7. The chairman shall nominate another member of the Financial Panel temporarily to act in his stead on such occasions or for such periods when the chairman may for any reason be unable to carry out his duties.
8. In the event that the chairman shall for any reason be unable to carry out his duties as chairman, the parties shall agree upon the identity of his replacement. In the absence of such agreement, or in the absence of agreement between the parties as to the identity of the initial chairman, the other members of the Financial Panel shall by means of a majority vote choose one of their number to act as chairman. Until such time as a chairman or replacement chairman has been agreed or appointed, the eldest member of the Financial Panel shall act as such.

Schedule 5 Appendix 2 Part II   Composition of the technical panel

1. The Technical Panel shall consist of 5 members including the chairman.
2. The Secretary of State and the Concessionaire shall endeavour in good faith and with due expedition to agree the identity of persons willing and suitable to

act as the initial members of the Technical Panel, and of one such person to act as its chairman, and of willing and suitable replacements whenever the need therefor arises.

3. In the event that the Secretary of State and the Concessionaire are unable to agree any or all of the initial members of the Technical Panel within 14 days of execution of this Concession Agreement, either party may apply to the nominating authority identified below to appoint such person or persons. Such party shall notify the nominating authority of the identity of those persons who have been considered by the parties under paragraph 2 above and who have been thought by either party not to be suitable. In the absence of agreement to the contrary of both parties the nominating authority shall not nominate any person who shall have been so identified.

4. The nominating authority for the purpose of appointing members of the Technical Panel shall be the President for the time being of the Institution of Civil Engineers or in his absence the Vice-President for the time being of the said institution.

5. With a view to facilitating the nominating authority's task of identifying suitable candidates for appointment, either party may forward to the nominating authority any relevant description which has been agreed by the parties of the likely background and experience of a suitable appointee and/or of the likely issues which the appointee as a member of the Technical Panel would be required to decide. The nominating authority may have regard to such description but shall not be bound by its terms. Subject to any such description, the nominating authority may have regard to the parties' present intention that the members of the Technical Panel should, so far as is reasonably practicable, possess between them expertise in the following particular areas:

   5.1 Superstructure
      (i)   Suspension and cable stayed bridges
         (a)   Design queries including non linear behaviour
         (b)   Fatigue in orthotropic steel decks, cables and connections
         (c)   Wing loading and aerodynamic behaviour.
      (ii) Approach spans
         (a)   Precast prestressed concrete design and construction
         (b)   Unbonded tendons
         (c)   Half joints
         (d)   Elastomeric bearings.
   5.2 Sub-Structure and Foundations
      (i)   Ground Conditions
         (a)   Geotechnics
         (b)   Rock mechanics.
      (ii) Estuarial Conditions
         (a)   Tide, wave and impact loading
         (b)   Local knowledge
   5.3 General
      (i)   Durability
         (a)   Steel
         (b)   Concrete
         (c)   Surfacing
      (ii) Environment
         (a)   Assessment of effects

6.   The chairman shall nominate another member of the Technical Panel temporarily to act in his stead on such occasions or for such periods when the chairman may for any reason be unable to carry out his duties.

7.   In the event that the chairman shall for any reason be unable to carry out his duties as chairman, the parties shall agree upon the identity of his replacement. In the absence of such agreement, or in the absence of agreement between the parties as to the identity of the initial chairman, the other members of the Technical Panel shall by means of a majority vote choose one of their number to act as chairman. Until such time as a chairman or replacement chairman has been agreed or appointed, the eldest member of the Technical Panel shall act as such.

Appendix 5
The Woolf Reforms and the Civil Procedure Rules

Concerns over the future development of civil justice led to the appointment of a senior member of the judiciary, Lord Woolf, to preside over the preparation of two reports, *Access to Justice*, in which he set out his blueprint for change. He identified the requirements of civil justice as:

- fairness
- respect for the litigants
- cost effectiveness
- reasonable speed
- understandability
- responsiveness
- provision of early certainty to litigants
- adequate resources, manpower and information technology.

The problems of civil litigation were obvious:

- costs frequently exceeded the value of the claim
- slowness
- an unequal struggle between the impecunious litigant and the big organisation which could throw lawyers at a problem
- difficulties in forecasting the outcome
- impenetrability for most litigants, however well informed.

Lord Woolf proposed a new system based on:

- avoidance of litigation, wherever possible, with greater emphasis on mediation and ADR generally in civil disputes
- litigation to be less adversarial with the parties more co-operative. Excessive nit-picking which fails to take cases forward was to be avoided
- less complex procedures
- quicker progress of cases to trial
- litigation to be more affordable, more predictable with the costs more related to the value of the claim
- greater equality between litigants
- courts to be run to serve litigants
- judges to play a greater role in case management.

To declare war on the old Rules of the Supreme Court and the County Court Rules, sometimes called the White and Green Books, the Civil Procedure Act 1997 received the Royal Assent on 27 February 1997. Section 1 of the Act provided that the practice and procedure of the Civil Courts was to be determined by Civil Procedure Rules. Section 2 allowed for the establishment of a Rules Committee with section 3 providing for rules to be made by Statutory Instrument. Section 4 allowed the Lord Chancellor to amend, repeal or revoke any enactment necessary or desirable to give effect to the new Civil Procedure Rules. A Civil Procedure Modification of Enactments Order was approved in 1998.

The new Civil Procedure Rules came into force on 27 April 1999. Where possible, apart from modernising the language of the old White Book and the Green Book, the rules of the Supreme Court and the County Court rules have been consolidated into one rule with appropriate amendments. The rules are divided into Parts which, in turn, are subdivided into rules and further subdivided into paragraphs and subparagraphs. Part one of the rules contains the Overriding Objective. The new procedural code is intended to be an exhaustive (no more lawyer skirmishes on what they mean) although of course, precedence will, in time, be set in the courts. Rule 1.1(ii) defines what is meant by dealing with cases justly:

- ensuring that the parties are on an equal footing
- saving expense
- dealing with cases in ways which are proportionate:

    o to the amount of money involved
    o to the importance of the case
    o to the complexity of the issues, and
    o to the financial position of the parties.

- ensuring that cases are dealt with expeditiously and fairly
- allotting to cases an appropriate share of the court's resources or taking into account the need to allocate resources to other cases.

Under rule 1.3 the parties have an obligation to assist the court to achieve the Overriding Objective. The rules are based on 'a fundamental transferring of responsibility for the management of civil litigation from litigants and their legal advisors to the Courts'.

Cases management is not new and lawyers had a dry run with the *Practice Direction (Civil Litigation: Case Management)* [1995] 1 AER 385, announced by a former Lord Chief Justice and the Vice Chancellor on 24 January 1995.

A cornerstone of the new system is the early establishment of trial dates. Traditionally cases have been listed with estimated lengths of hearing based on the whimsical 'say so' of counsel. Now, in principle, under the Woolf reforms the parties will be better prepared even before the proceedings get under way. However, existing experience shows that nothing focuses minds better than the imminent approach of a trial date. Shorten the periods to trial, and potentially the greater the enthusiasm for settlement.

The main features of the reforms are:

- Proceedings will commence on a standard form to be used both in High Court and County Court actions. The fully pleaded case may be included in the claim form or provided as a separate document. It will have to be endorsed with a Statement of Truth to accept responsibility for the truthfulness of the document and may be used as evidence both in interlocutory matters (summary judgment, interim payments, etc.) or at trial. No claim under £15,000 will be commenced in the High Court. Under rule 7.1 proceedings start when the court issues the claim form, and this date is obviously of significance for limitation questions. Part 7 and Part 16 deal with the contents of the claim form. The form contains a concise statement of the nature of the claim, including remedies sought, but can go beyond a traditional pleading and contain evidence. The claim form must be served within four months of issue or six months if the defendant is outside the jurisdiction of the courts of England and Wales. Particular rules relate to the service of a defence. Subject to some slight modifications, the principle of summary judgement is retained. This is dealt with in Part 24. In accordance with the modernising theme Further and Better Particulars will be replaced by a Request for Further Information. Affidavits will largely disappear to be replaced in most applications by witness statements, all subject to the Statement of Truth. The old rules in RSC Order 24 and County Court Rules Order 14 on discovery are now replaced by new rules set out in Part 31. They require the discovery (disclosure) of documents which are or have been in the possession, custody or power of a party and which relate to the matters in question in the action. Discovery or disclosure relates to:

- a party's own documents
- adverse documents
- relevant documents
- train of enquiry documents, being the category which has led to most of the problems since.

The intention of the new rules is to reduce the cost of discovery by controlling and reducing the enthusiasm of many solicitors for the photocopier. There will be a separate Practice Direction. Under rule 31.5 disclosure only takes place upon an order of the Court with no party required to give disclosure voluntarily or automatically. Under paragraph 3 of rule 31.5, although the parties may limit discovery or dispense with it altogether they cannot extend it beyond that which is included in rule 31. The principles of standard disclosure as set out in rule 31.5(1a) will be the exclusive mechanism for the fast track, with additional discovery in the multi track dealt with in rule 31.12. Under rule 31.6(2) where a party considers that particular aspects of discovery would be disproportionate to the issues in the case, disclosure is not necessary although a statement needs to be issued that disclosure has been withheld on the basis of disproportionality. Of further importance is the disclosure statement which is referred to in rule 31.10(6) which indicates the extent of the search that has been made certifying that the party understands the duty to disclose and that to the best of his knowledge he has carried out that duty.

Of particular significance for technical disputes are the rules on experts to be found in Part 35. Under rule 35.3 the expert has an overriding duty to the court and is obliged to help the court on matters within his expertise. Rule 35.3(2) states that his duty overrides any obligation that he owes to his client. Experts cannot be called to give evidence without the court's permission; read rule 35.4(4) carefully. The court may limit in advance the expert's fees and expenses that may be recovered from the other party. This may lead to applications by a party to limit the costs in the event that they lose. To limit further the costs effect under rule 35.5(2) in fast track claims (see below) the court will never direct the expert to attend a hearing to give oral evidence. Under rule 35.6 parties can put to experts written questions about his report. Rule 35.7 allows for a single expert.

Other procedures are included to tidy up what Lord Woolf identified as the problems with expert witnesses. Under rule 35.10 the substance of all material instructions, whether written or oral, upon which the report was based must be expressed. Those instructions are not privileged and will ordinarily be disclosed. Under rule 35.11 reports which have been disclosed, but not used, can be used by any other party and under 35.12 the question of without prejudice meetings of experts has been addressed. The court is now to take complete control of expert meetings to avoid solicitors fettering the expert. Under rule 35.14 the expert can ask the court for guidance.

On costs, the principal change is that the recoverable costs will be subject to a test of proportionality. Costs will be assessed by reference to what was sensible and reasonable in all the circumstances. Detailed principles are set out in rule 43.

Court users must distinguish fast track from multi track, a central pillar of case management. Ordinarily the fast track will cover claims up to £15,000 with multitrack claims over £15,000. Fast track will also affect cases not lasting longer than one day. If no value is attached to a claim or there are other relevant factors, the court will, in allocating a case consider:

- amounts in dispute and the remedies sought
- complexity
- number of parties
- counterclaims
- amount of oral evidence
- public importance
- the view of the parties
- the circumstances of the parties
- the detailed provisions are found in Parts 28 (fast track) and Parts 29 (multi track).

Under the CPR, ADR plays a much more prominent role. According to rule 1.4(2)(e) active case management includes:

'Encouraging the parties to use an alternative dispute resolution procedure if the court considers that appropriate and facilitating the use of such procedure.'

This all ties in with the CPR being subject in Part 1 rule 1.1(1) to the overriding objective requiring the courts to 'deal with cases justly'. ADR is expressly provided for in Part 26, rule 26.4. Under rule 26.4(1):

'A party may, when filing the completed allocation questionnaire, make a written request for the proceedings to be stayed while the parties try to settle the case by alternative dispute resolution or other means.'

Perhaps more controversially, under rule 26.4(2):

'Where –
(a) all parties request a stay under paragraph (1); or
(b) the court, of its own initiative, considers that such a stay would be appropriate, the court will direct that the proceedings be stayed for one month.'

The court may extend the stay 'until such date or for such specified period as it considers appropriate' (rule 26.4(3)).

If parties do not follow ADR, the question arises what sanctions there may be. There are new costs' provisions in the CPR. Under rule 44.5(3), the court must have regard, when assessing the amount of costs, to a number of factors including:

'(a) the conduct of all the parties, including in particular –
  (i) conduct before, as well as during, the proceedings; and
  (ii) the efforts made, if any, before and during the proceedings in order to try to resolve the dispute.'

Obviously the CPR are new and time will tell how they are developed and deployed by the courts.

Some see the Woolf Reforms and the new CPR as a breath of fresh air, a sensible way forward which will curtail the excesses of the legal system. For cynics, lawyers will always find ways of manipulating the system.

References

Chapter 1

1. Brown and Marriott, *ADR Principles and Practice* (1993) at page 5. Sweet and Maxwell, London.
2. for example, Acland, A.F., *A Sudden Outbreak of Common Sense: Managing Conflict through Mediation* (1990) at page 69. Century, London.
3. Allen, Richard K. *Dispute Avoidance and Resolution for Consulting Engineers* (1993) at page 18. ASCE Press, New York.
4. Clegg, Stewart R., Contracts Cause Conflicts, page 143, in *Construction Conflict Management and Resolution* (1992) edited by Fenn and Gameson. E & FN Spon, London.
5. Groton, James P., *Alternative Dispute Resolution in the Construction Industry* (1993) cumulative supplement at page 11. John Wiley & Sons, New York.
6. Latham, M., *Constructing the Team*, final report of the Joint Government Industry Review of Procurement and Contractual Arrangements in the United Kingdom (1994) in para 5.5.
7. Allen, *op. cit.*, at page 11.
8. Timpson, John, *The Architect in Dispute Resolution* (1994) at page 13. RIBA Publications, London.
9. Baden Hellard, R., Construction Conflict – Management and Resolution, page 39, in *Construction Conflict Management and Resolution* (1992) edited by Fenn and Gameson. E & FN Spon, London.
10. Brandon, P., Hibberd, P., Basden, A., Kirkham, J. & Tetlow, S., *Intelligent Authoring of Construction Contracts (INCA)*, final report (1993) of SERC Funded Research Project GR/G200011, the University of Salford and the University of Glamorgan.
11. Latham *op. cit.*, at page vii, item 9 of executive summary.
12. Allen *op. cit.*, at page 25.
13. McCanlis, E.W., *Tendering and Contractual Arrangements* (1967) Research and Information Group of the Quantity Surveyors' Committee, Royal Institution of Chartered Surveyors.
14. Timpson *op. cit.*, at page 13.
15. Lavers, A., Construction Conflict – Management and Resolution Analysis and Solutions, page 7, in *Construction Conflict Management and Resolution* (1992) edited by Fenn and Gameson. E & FN Spon, London.
16. Smith, M.C.G., Facing up to Conflict in Construction, page 28, in *Construction Conflict Management and Resolution* (1992) edited by Fenn and Gameson. E & FN Spon, London.
17. Baden Hellard, *op. cit.*, at page 35.
18. Langford, D., Kennedy, P., and Sommerville, J., Contingency Management of Conflict: Analysis of Contract Interfaces, at page 64, in *Construction Conflict Management and Resolution* (1992) edited by Fenn and Gameson. E & FN Spon, London.

19. Harding, D., Building Without Conflict, *Building,* November 1991.
20. for example, Hibberd, P., *Building Contract – Variations* (1980) MSc Thesis, UMIST; Bromilow, F.J., Contract Time Performance, Expectations and the Reality (1969) *Building Forum,* **1** (3) and The Nature and Extent of Variations in Building Contracts (1970), *Building Economist,* **9** (3); Watts, V. & Scrivener, J., Review of Australian Building Disputes Settled by Litigation, at pages 209–18, in *Construction Conflict Management and Resolution* (1992) edited by Fenn and Gameson. E & FN Spon, London.
21. Newey, Judge John, The Construction Industry, at page 22, in *Construction Conflict Management and Resolution* (1992) edited by Fenn and Gameson. E & FN Spon, London.
22. McGivering, I.C., Conflict, pages 94–5, in *A Handbook of Management* (1983) edited by Kempner. Penguin Books, Hardmondsworth.
23. Smith *op cit.,* at page 8.
24. Rahim, M.A., Managing Conflict in Organisations, at pages 369–77, in *Construction Conflict Management and Resolution* (1992) edited by Fenn and Gameson. E & FN Spon, London.

Chapter 2
1. Robert Coulson, *Professional Mediation of Civil Disputes* (1984), at pages 6–7. American Arbitration Association.
2. *Interim Report to the Lord Chancellor on the Civil Justice System in England and Wales,* Lord Woolf, June 1995.
3. Quoted by Lord Alexander of Weedon QC in *Training Lawyers – Healers or Hired Guns?* Child & Co. lecture, 15 March 1995, at page 3.
4. Quoted by Lord Alexander of Weedon QC, *ibid.,* at page 5.
5. Survey by the law firm, Herbert Smith, *The Times,* January 1995.
6. *Building,* 7 February 1992, at page 9.
7. [1984] 1 QB 644, at page 70.
8. National Building and Construction Council Joint Working Party (1989) *Strategies for the Reduction of Claims and Disputes in the Construction Industry* – a research report (various authors). NBCC, Canberra.
9. Quoted by Lord Alexander of Weedon QC, *op. cit.,* at page 21.
10. *The Lawyer,* 13 June 1995, at page 15.
11. *Construction Management and Economics,* **15** (6), Brooker and Lavers, at pages 519–26.
12. December 1993.
13. July 1994 HMSO, London.
14. *Report of the Committee on the Placing and Management of Contracts for Building and Civil Engineering Work,* chaired by Sir Harold Banwell, 1964, HMSO, London.
15. *The Times,* Frances Gibb, 27 August 1996.
16. Resolving Disputes Without Going to Court, Lord Chancellor's Department, London, 1995.
17. *A Consultation Paper on Draft Clauses and Schedules of an Arbitration Bill,* Department of Trade and Industry, February 1994.
18. *Draft Clauses of an Arbitration Bill; Consultative Paper on an Arbitration Bill,* Department of Trade and Industry, July 1995.
19. Arbitration Bill 1996, HL Bill 60.
20. Construction Industry Model Arbitration Rules (CIMAR), February 1998, Society of Construction Arbitrators.

21. Section 34(2)(d), Arbitration Act 1996.
22. Section 34(2)(f), Arbitration Act 1996.
23. Section 34(2)(g), Arbitration Act 1996.
24. In *Gilbert Ash (Northern) Limited* v. *Modern Engineering (Bristol) Limited* (1973) AER 195 at page 223.
25. Section 39(2)(a), Arbitration Act 1996.
26. See *Phillip Alexander Securities and Futures Limited* v. *Bamberger and Others; Same* v. *Gilhaus, The Times*, 22 July 1996.

Chapter 3

1. *Members Handbook*, British Academy of Experts, 1992, at page A1/1.
2. *The Lawyer*, 13 June 1995, at page 15.
3. Lavers & Brooker, Perceptions of alternative dispute resolution as constraints upon its use in the UK construction industry, *Construction Management and Economics*, **15** (6), November 1997, at pages 519–26.
4. *JCT Practice Note 28*, Mediation in a Building Contract or Sub-Contract Dispute, July 1995, RIBA.
5. In *Allco Steel (Queensland) Pty Ltd* v. *Torres Strait Gold Pty Ltd* (1990) Supreme Court of Queensland, unreported, 12 March 1990.
6. See also comments of Rogers, J. in *AWA Limited* v. *Daniels*, unreported, 24 February 1992.
7. Compare with *Coal Cliff Collieries Pty Ltd* v. *Sijehama Pty Ltd* [1991] 24 NSWLR 1 where an 'agreement to negotiate' was held to be unenforceable.
8. See Boulle, case note on *AWA Limited* v. *Daniels*, Vol 3 ADR Journal 272 at page 275.
9. The Commercial Arbitration Act applies uniformly in all states except Queensland and the amendment has been, or is anticipated to be, adopted in all participating states.
10. *Seeking Harmony*, M. Scott Donahey (1995) 61 JCI Arb. 4, at page 279.
11. *ibid.*, at page 280.
12. Arbitration (Amendment) Ord. No. 10/82.
13. *Hong Kong Dispute Solutions*, Hong Kong International Arbitration Centre.
14. *ibid.*, Chapter 4, at page 10.
15. The Law Reform Commission of Hong Kong, *Report on Commercial Arbitration (Topic 1)*, paras 10.25–10.31.
16. The Law Reform Commission of Hong Kong, *Report on the Adoption of the UNCITRAL Model Law of Arbitration (Topic 17)*, 1987, paras 4.32–4.35.
17. Chapter 341, *Arbitration Ordinance*, 1989.
18. Report and Recommendations of The Chief Justice's Committee on The Desirability of Introducing a Court Annexed Mediation Scheme in Hong Kong and Related Matters, August 1993.
19. Alternative Dispute Resolution Case Management, *Hong Kong Lawyer*, Phillip Wright, September 1994.
20. 8 July 1992.
21. Government of Hong Kong General Conditions of Contract for Civil Engineering Works, 1990 and 1993 Editions.
22. 2 May 1989.
23. Hong Kong International Arbitration Centre, July 1992.
24. Dispute Resolution in the New Hong Kong International Airport Core Programme Project – Part 2, Dean Lewis (1994) *International Construction Law Review*, part 1, at pages 27–9.
25. *ibid.*, at page 27.

26. *ibid.*, at page 28.
27. Dispute Resolution in the New Hong Kong International Airport Core Programme Project – Part 3, Dean Lewis (1995) *International Construction Law Review*, part 1, at pages 131–6.
28. *ibid.*, at page 132.
29. July 1994 revision.
30. Construction Industry Council (CIC) (1994) report *Dispute Resolution*, which identifies the disputes that arise in the construction industry and existing methods of resolution.

Chapter 4
1. Construction Industry Council (CIC) (1994) report *Dispute Resolution*, which identifies the disputes that arise in the construction industry and existing methods of resolution.
2. Brown and Marriott, *ADR Principles and Practice* (1993), at page 19. Sweet and Maxwell, London.
3. National Joint Consultative Committee for Building, *Alternative Dispute Resolution*, Guidance Note 7, Appendix 2 (1993).
4. Hollands, D., Amicable Dispute Resolution (ADR) Procedures, *International Construction Law Review* (1990) **7**, part 4, at pages 452–62.
5. Wright, M., Mediation – The Idea, Influences on the development of conflict resolution, *Justice of the Peace & Local Government Law*, 27 November 1993.
6. Naughton, P., Alternative Forms of Dispute Resolution, at page 197, *Construction Law Journal* (1990) **6**, part 3.
7. Hollands, *op. cit.*
8. Fisher, R. & Ury, W., *Getting to Yes – Negotiating Agreements Without Giving In*, 1991, Business Books Ltd, London.
9. Feinberg, Kenneth R., *A Procedure for the Voluntary Mediation of Disputes*, Kaye Scholer, Fierman, Hays and Handler, New York (1990), also given as a paper to IBA Business Law Section Hong Kong 1991. (In exceptional circumstances there may be a conference-call mediation. A landlord and tenant dispute where the parties were in Texas and San Mateo, California, respectively was subject to mediation by telephone: *Consensus*, April 1996, No. 30, at page 9, MIT-Harvard Public Disputes Program.)
10. Selig, Louis (United States Arbitration and Mediation) reported by Hilary Heilbron in the *Gazette* 90/44, 1 December 1993.
11. Quick, R.W., Costs in Arbitration Proceedings, at pages 227–8, in *Construction Conflict Management and Resolution* (1992) edited by Fenn and Gameson. E & FN Spon, London.
12. Australian Federation of Construction Contractors, *Strategies for the Reduction of Claims Disputes in the Construction Industry* – a research report (1989).
13. Hollands, *op. cit.*, at page 453.
14. Allen, R.K., *Dispute Avoidance and Resolution for Consulting Engineers* (1993), at page 61, ASCE Press, New York.
15. *Amicable Dispute Settlement* (1993) at page 5, Construction Disputes Resolution Group.
16. Naughton, P., Alternative Forms of Dispute Resolution, at page 199, *Construction Law Journal* (1990) **6**, part 3.
17. Williams, R., Alternative Dispute Resolution (ADR): Salvation or Chimera, page 106, *Arbitration*, May 1990.
18. Gaede, A.H. ADR – The US Experience and Some Suggestions for International

Arbitration: The Observations of an American Lawyer, at page 21, in *The International Construction Law Review* (1991) **8**, part 1.

19. Wright, *op. cit.*.

20. Fisher, R., Ury & Patton, *Getting to Yes*, second edition (1991), at page 29, Business Books, London.

21. *ICC Rules of Conciliation*, International Chamber of Commerce, Paris, France.

Chapter 5

1. *New Civil Engineer*, 7/14 April 1994.

2. Latham, M., *Constructing the Team*, final report of the Joint Government Industry Review of Procurement and Contractual Arrangements in the United Kingdom Construction Industry (1994) in paras 9.4–9.7 and 9.14.

3. Fair Construction Contracts Consultation Document, DoE, May 1995.

4. Department of Environment News Release, Notes to Editors, paragraph 2, 20 November 1995, at page 2.

5. Adjudication is not Arbitration by another name, *CIC Digest*, Issue 28, 22 March 1996.

6. Latham, M., Interim Report, Trust and Money (1993) and *Constructing the Team*, final report of the Joint Government Industry Review of Procurement and Contractual Arrangements in the United Kingdom (1994).

7. Ron Denny, Former Deputy Director of the British Property Federation in an interview in *Construction News*, 13 June 1991, as referred to in *Adjudicators, Experts and Keeping out of Court*, Mark C. McGaw, *Current Developments in Construction Law*, Fourth Annual Construction Conference, King's College, London, 20 September 1991.

8. *Wiseman* v. *Borneman* [1971] AC 297, at page 308.

9. *Russell* v. *Duke of Norfolk* [1949] 1 AER 109, at page 118.

10. In Mustill and Boyd, *Commercial Arbitration* (1989), second edition, at page 41. Butterworths, London.

11. Bernstein, R. and Wood, D. *Handbook of Arbitration Practice* (1993), second edition, at pages 13 and 14, Sweet & Maxwell, London.

12. *Commercial Arbitration*, second edition (1989) at page 49, Butterworths, London.

13. *ibid.*, at page 49.

14. *Hickman & Co.* v. *Roberts and Others* [1913] AC 229, at page 234.

15. *Chambers* v. *Goldthorpe* [1901] 1 KB 624, at page 636.

16. *Sutcliffe* v. *Thackrah* [1974] AC 727, at pages 757 and 758.

17. *Arenson* v. *Casson Beckman Rutley & Co.* [1977] AC 405, at page 424.

18. *ibid.*, at page 428.

19. *ibid.*, at page 440.

20. *Ponsarn Investments Limited* v. *Kansallis-Osake-Pankki* [1992] 1 EGLR, at page 167.

21. *ibid.*, at page 169.

22. *Dixons Group plc* v. *Jan Andrew Murray-Obsynski* (1998) 86 BLR 16, at page 30.

23. *Northern Regional Health Authority* v. *Derek Crouch Construction Company Ltd* [1984] 1 QB 644, at page 664.

24. *Tubeworkers Ltd* v. *Construction Ltd* [1985] 30 BLR, at pages 77 and 78.

Chapter 6

1. Latham, M., *Constructing the Team*, final report of the Joint Government Industry Review of Procurement and Contractual Arrangements in the United Kingdom Construction Industry (1994).

2. Construction Industry Council (CIC) (1994) report *Dispute Resolution*, which

identifies the disputes that arise in the construction industry and existing methods of resolution.

3. Dispute Review Boards, *ADR A Practical Guide to Resolve Construction Disputes* (1994) John R. Kohnke, at pages 267–8, American Arbitration Association.

4. Dispute Review Boards, Burt Campbell (1994) 60 JCI Arb 1, at pages 17–18.

5. Kenneth Severn, Arbitrator and Consulting Engineer.

6. Kenneth Severn, Arbitrator and Consulting Engineer.

7. *Amicable Dispute Settlement*, Construction Disputes Resolution Group (1994) CDRG Mediation and Arbitration Services, Kenneth Severn.

8. Dispute Review Boards and Adjudicators, *International Construction Law Review* (1993) April, at pages 157–71.

9. *ibid.*, at page 159.

10. *World Bank Sample Bidding Document for the Procurement of Works*, December 1990. See also *World Bank Standard Bidding Documents for the Procurement of Works*, January 1995. Section 13 sets out a mandatory procedure to be used by three member dispute review boards for contracts of more than $50m. Smaller contracts, between $10m and $50m, may have either the three member dispute review board or a single dispute review expert. The *'Orange Book' Conditions of Contract for Design–Build and Turnkey* (1995) provides for a dispute adjudication board. The *Orange Book* states that 'decisions' will be given effect to forthwith unless and until revised.

11. *A Claims Review Board as a Way for Amicable Settlement of Disputes and Other Considerations on the Subject of Claims* [1986] ICLR 498, G. Lodigiani.

12. *Adjudication as operated on the construction of the Dartford River Crossing (The Queen Elizabeth II Bridge)* (1994), 60 JCI Arb. 1, at pages 13–6, Michael E. Morris.

13. *ibid.*, at page 13.

Chapter 7

1. *The Manager's Guide to Resolving Legal Disputes – Better Results Without Litigation*, James F. Henry and Jethro K. Lieberman (1985), Chapter 3, at pages 19–25. Harper & Row, New York.

2. *Mini-Trial*, Douglas H. Yarn, *ADR A Practical Guide to Resolve Construction Disputes* (1994), at page 234, American Arbitration Association.

3. *Guidelines for Supervised Settlement Procedure*, ('Mini-Trial'), (1990), The Chartered Institute of Arbitrators, London.

4. Revised November 1990.

5. *Preventing and Resolving Construction Disputes* (1991), at pages A/1–A/8, Center for Public Resources, Inc., New York.

6. *Quiet revolution brews for settling disputes*, Rob McManamy, ENR 26 August 1991, at page 22.

7. Yarn Douglas H., MedArb, *ADR A Practical Guide to Resolve Construction Disputes* (1994), at page 217, American Arbitration Association.

8. *ibid.*, at page 225.

9. *Re: Catalina (Owners) and Norma MV (Owners)* [1938] 61 Lloyds LR 360.

10. *Re: Elliot ex parte South Devon Railway Company* (1848) 12 Jur 445.

11. *R. v. Gough*, The Times, 24 May 1993.

12. *The Elissar* [1984] 2 Lloyds Rep 84.

13. *Banque Keyser Ullman v. Skandia Insurance* [1991] 2AC 249, at page 280, per Lord Templeman; *Upjohn v. Oswald; Vernon v. Bosley Construction*, both unreported but to be found in *Construction Law Yearbook 1996*, John Wiley, at page 199.

14. Section 34.

15. Unfair Contract Terms Act 1977.
16. Mustill and Boyd, *Commercial Arbitration* (1989) second edition, at page 283, Butterworth, London.
17. *Haigh* v. *Haigh* (1861) 31 LJ Ch 420, per Turner LJ.
18. *Med-Arb – Can it Work?* (1994) 60 JCI Arb 1, at pages 1 and 2.
19. *ibid.*, at page 3.
20. *The Med-Arb debate continued* (1995) JCI Arb 2, at pages 111 and 112.
21. 15 July 1993.
22. These are usefully summarised in *Corporate Counsel's Primer on Alternative Dispute Resolution Techniques* (1990), William A. Hancock, Chapter 401, Business Laws Inc., Chesterland, Ohio.
23. *ibid.*, para 401.019.
24. MEDALOA^{TM}, *Mediation and Last Offer Arbitration*, American Arbitration Association.
25. *ibid.*, at page 6.

Chapter 8

1. A notable example of litigation which backfired was Mowlem's *Carlton Gate* claim. According to *Contract Journal*, 10 August 1995: 'What is amazing is that Mowlem continued its battle even after declaring a £123 million loss for 1993. On what quality of legal advice? Some might ask.' One lawyer speculated: 'You always try to give the client some good news in the early days. Perhaps that's what ... did, and a bandwagon was set up'.
2. Matthew Chapter 5, verse 9.
3. *Tai Hing Cotton Mill Limited* v. *Liu Chong Hing Bank Limited* [1986] AC 80 (PC).
4. *Lancashire & Cheshire Associations of Baptist Churches Inc.* v. *Howard & Seddon Partnership* [1993] 3 AER 467; *Conway* v. *Crowe Kelsey Partner & Another* (1994) CILL 927; *Wessex Regional Health Authority* v. *HLM Design Limited* (1993) CILL 907; *Holt and Another* v. *Payne Skillington (a Firm) and Another*, The Times, 22 December 1995 (CA).
5. [1927] WN 290.
6. *Rush and Tompkins Ltd* v. *Greater London Council* [1989] AC 1280, at page 1299.
7. *La Roche* v. *Armstrong* [1922] 1 KB 485.
8. *D* v. *NSPCC* [1978] AC 171, at pages 236 and 237, Lord Simon.
9. *Dispute Resolution Times*, New York, Spring 1996.
10. *Northern Regional Health Authority* v. *Derek Crouch Construction Company Ltd* [1984] 1 QB 644, at page 664, per Dunn LJ.
11. *Courtney and Fairburn Ltd* v. *Tolnini Brothers (Hotels) Ltd* [1975] 1 WLR 297, at pages 301 and 302.
12. [1992] AC 128, at pages 181–2.
13. *Pitt* v. *PHH Assc and Management Ltd* [1994] 1 WLR 327, at pages 332–3.

Chapter 10

1. see Clause 66(1)(2)(3) of ICE Conditions of Contract 6th Edition.
2. Parris, J., *Default by Sub-Contractors and Suppliers* (1985) at page 139, Collins, London.

Chapter 12

1. Expert Procedures – Model Forms of Conditions of Contract for Process Plants, IChemE, February 1998, pages 7–8.

2. Expert Procedures – Model Forms of Conditions of Contract for Process Plants, IChemE, February 1998.

Chapter 13
1. Consultative Document, The New Engineering Contract, Guidance Notes, page 60, Thomas Telford, London.

Table of Cases

Index